2024

ICCPC®

INTERNATIONAL CODE COUNCIL **PERFORMANCE** CODE®

FOR BUILDINGS AND FACILITIES

AF853349

ICC
INTERNATIONAL CODE COUNCIL®

2024 International Code Council Performance Code® for Buildings and Facilities

First Printing: August 2023

ISBN: 978-1-959851-89-9 (soft-cover edition)
ISBN: 978-1-959851-90-5 (PDF download)

COPYRIGHT © 2023
by
INTERNATIONAL CODE COUNCIL, INC.

ALL RIGHTS RESERVED. This 2024 *International Code Council Performance Code®* for *Buildings and Facilities* is a copyrighted work owned by the International Code Council, Inc. ("ICC"). Without separate written permission from the copyright owner, no part of this publication may be reproduced, distributed or transmitted in any form or by any means, including, without limitation, electronic, optical or mechanical means (by way of example, and not limitation, photocopying or recording by or in an information storage and/or retrieval system). For information on use rights and permissions, please contact: ICC Publications, 4051 Flossmoor Road, Country Club Hills, Illinois 60478; 1-888-ICC-SAFE (422-7233); https://www.iccsafe.org/about/periodicals-and-newsroom/icc-logo-license/.

Trademarks: "International Code Council," the "International Code Council" logo, "ICC," the "ICC" logo, "International Code Council Performance Code," "ICCPC" and other names and trademarks appearing in this publication are trademarks of the International Code Council, Inc., and/or its licensors (as applicable), and may not be used without permission.

T029304

PRINTED IN THE USA

NEW DESIGN FOR THE 2024 INTERNATIONAL CODES

ICC INTERNATIONAL CODE COUNCIL®

IBC IRC IFC IPC IMC IECC IEBC IFGC IPMC IPSDC IWUIC IZC ICCPC IgCC ISPSC

The 2024 International Codes® (I-Codes®) have undergone substantial formatting changes as part of the digital transformation strategy of the International Code Council® (ICC®) to improve the user experience. The resulting product better aligns the print and PDF versions of the I-Codes with the ICC's Digital Codes® content.

The changes, promoting a cleaner, more modern look and enhancing readability and sustainability, include:

Streamlined lists

QR codes to identify code changes more accurately
(For further details, see Formatting Changes to the 2024 International Codes.)

Consistent grouping of associated content
(e.g., tables immediately follow parent sections)

2024

Single-column text

Shading for table headers and notes

Modernized font styles
A B C D
a b c d

More information can be found at iccsafe.org/design-updates.

PREFACE

FORMATTING CHANGES TO THE 2024 INTERNATIONAL CODES

The 2024 International Codes® (I-Codes®) have undergone substantial formatting changes as part of the digital transformation strategy of the International Code Council® (ICC®) to improve the user experience. The resulting product better aligns the print and PDF versions of the I-Codes with the ICC's Digital Code content. Additional information can be found at iccsafe.org/design-updates.

Replacement of Marginal Markings with QR Codes

Through 2021, print editions of the I-Codes identified technical changes from prior code cycles with marginal markings [solid vertical lines for new text, deletion arrows (➡), asterisks for relocations (★)]. The 2024 I-Code print editions replace the marginal markings with QR codes to identify code changes more precisely.

A QR code is placed at the beginning of any section that has undergone technical revision. If there is no QR code, there are no technical changes to that section.

In the following example from the 2024 *International Code Council Performance Code® for Buildings and Facilities* (ICCPC®), a QR code indicates there are changes to Section 101 from the 2021 ICCPC. Note that the change may occur in the main section or in one or more subsections of the main section.

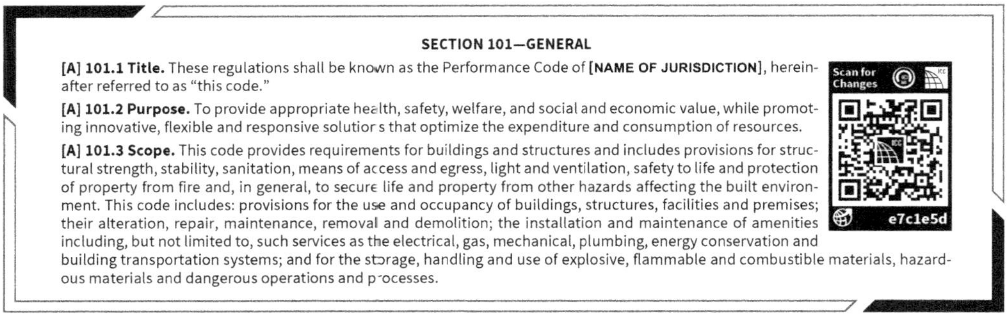

SECTION 101—GENERAL

[A] 101.1 Title. These regulations shall be known as the Performance Code of **[NAME OF JURISDICTION]**, hereinafter referred to as "this code."

[A] 101.2 Purpose. To provide appropriate health, safety, welfare, and social and economic value, while promoting innovative, flexible and responsive solutions that optimize the expenditure and consumption of resources.

[A] 101.3 Scope. This code provides requirements for buildings and structures and includes provisions for structural strength, stability, sanitation, means of access and egress, light and ventilation, safety to life and protection of property from fire and, in general, to secure life and property from other hazards affecting the built environment. This code includes: provisions for the use and occupancy of buildings, structures, facilities and premises; their alteration, repair, maintenance, removal and demolition; the installation and maintenance of amenities including, but not limited to, such services as the electrical, gas, mechanical, plumbing, energy conservation and building transportation systems; and for the storage, handling and use of explosive, flammable and combustible materials, hazardous materials and dangerous operations and processes.

To see the code changes, the user need only scan the QR code with a smart device. If scanning a QR code is not an option, changes can be accessed by entering the 7-digit code beneath the QR code at the end of the following URL: qr.iccsafe.org/ (in the above example, "qr.iccsafe.org/e7c1e5d"). Those viewing the code book via PDF can click on the QR code.

All methods take the user to the appropriate section on ICC's Digital Codes website, where technical changes from the prior cycle can be viewed. Digital Codes Premium subscribers who are logged in will be automatically directed to the Premium view. All other users will be directed to the Digital Codes Basic free view. Both views show new code language in blue text along with deletion arrows for deleted text and relocation markers for relocated text.

Digital Codes Premium offers additional ways to enhance code compliance research, including revision histories, commentary by code experts and an advanced search function. A full list of features can be found at codes.iccsafe.org/premium-features.

ACCESSING ADDITIONAL FEATURES VIA REGISTRATION OF BOOK

Beginning with the 2024 *International Mechanical Code®* (IMC®) and the 2024 *International Plumbing Code®* (IPC®), users will be able to validate the authenticity of their book and register it with the ICC to receive incentives. Digital Codes Premium (codes.iccsafe.org) provides advanced features and exclusive content to enhance code compliance. To validate and register, the user will tap the ICC tag (pictured here and located on the front cover) with a near-field communication (NFC) compatible device. Visit iccsafe.org/nfc for more information and troubleshooting tips regarding NFC tag technology.

ABOUT THE I-CODES

The 2024 I-Codes, published by the ICC, are 15 fully compatible titles intended to establish provisions that adequately protect public health, safety and welfare; that do not unnecessarily increase construction costs; that do not restrict the use of new materials, products or methods of construction; and that do not give preferential treatment to particular types or classes of materials, products or methods of construction.

The I-Codes are updated on a 3-year cycle to allow for new construction methods and technologies to be incorporated into the codes. Alternative materials, designs and methods not specifically addressed in the I-Code can be approved by the building official where the proposed materials, designs or methods comply with the intent of the provisions of the code.

The I-Codes are used as the basis of laws and regulations in communities across the US and in other countries. They are also used in a variety of nonregulatory settings, including:

- Voluntary compliance programs.
- The insurance industry.

- Certification and credentialing for building design, construction and safety professionals.
- Certification of building and construction-related products.
- Facilities management.
- "Best practices" benchmarks for designers and builders.
- College, university and professional school textbooks and curricula.
- Reference works related to building design and construction.

Code Development Process

The code development process regularly provides an international forum for building professionals to discuss requirements for building design, construction methods, safety, performance, technological advances and new products. Proposed changes to the I-Codes, submitted by code enforcement officials, industry representatives, design professionals and other interested parties, are deliberated through an open code development process in which all interested and affected parties may participate.

Openness, transparency, balance, due process and consensus are the guiding principles of both the ICC Code Development Process and OMB Circular A-119, which governs the federal government's use of private-sector standards. The ICC process is open to anyone without cost. Remote participation is available through cdpAccess®, the ICC's cloud-based app.

In order to ensure that organizations with a direct and material interest in the codes have a voice in the process, the ICC has developed partnerships with key industry segments that support the ICC's important public safety mission. Some code development committee members were nominated by the following industry partners and approved by the ICC Board:

- American Gas Association (AGA)
- American Institute of Architects (AIA)
- American Society of Plumbing Engineers (ASPE)
- International Association of Fire Chiefs (IAFC)
- National Association of Home Builders (NAHB)
- National Association of State Fire Marshals (NASFM)
- National Council of Structural Engineers Association (NCSEA)
- National Multifamily Housing Council (NMHC)
- Plumbing Heating and Cooling Contractors (PHCC)
- Pool and Hot Tub Alliance (PHTA) formerly The Association of Pool and Spa Professionals (APSP)

Code development committees evaluate and make recommendations regarding proposed changes to the codes. Their recommendations are then subject to public comment and council-wide votes. The ICC's governmental members—public safety officials who have no financial or business interest in the outcome—cast the final votes on proposed changes.

The I-Codes are subject to change through future code development cycles and by any governmental entity that enacts the code into law. For more information regarding the code development process, contact the Codes and Standards Development Department of the ICC at iccsafe.org/products-and-services/i-codes/code-development/.

While the I-Code development procedure is thorough and comprehensive, the ICC, its members and those participating in the development of the codes expressly disclaim any liability resulting from the publication or use of the I-Codes, or from compliance or noncompliance with their provisions. NO WARRANTY OF ANY KIND, IMPLIED, EXPRESSED OR STATUTORY, IS GIVEN WITH RESPECT TO THE I-CODES. The ICC does not have the power or authority to police or enforce compliance with the contents of the I-Codes.

Code Development Committee Responsibilities (Letter Designations in Front of Section Numbers)

In each cycle, proposed changes are considered by the Code Development Committee assigned to a specific code or subject matter. Committee Action Hearings result in recommendations regarding a proposal to the voting membership. Where changes to a code section are not considered by that code's own committee, the code section is preceded by a bracketed letter designation identifying a different committee. Bracketed letter designations for the I-Code committees are:

[A]	=	Administrative Code Development Committee
[BE]	=	IBC—Egress Code Development Committee
[BF]	=	IBC—Fire Safety Code Development Committee
[BG]	=	IBC—General Code Development Committee
[BS]	=	IBC—Structural Code Development Committee
[E]	=	Developed under the ICC's Standard Development Process
[EB]	=	International Existing Building Code Development Committee
[F]	=	International Fire Code Development Committee
[FG]	=	International Fuel Gas Code Development Committee
[M]	=	International Mechanical Code Development Committee

[P] = International Plumbing Code Development Committee

[SP] = International Swimming Pool and Spa Code Development Committee

For the development of the 2027 edition of the I-Codes, the ICC Board of Directors approved a standing motion from the Board Committee on the Long-Term Code Development Process to revise the code development cycle to incorporate two committee action hearings for each code group. This change expands the current process from two independent 1-year cycles to a single continuous 3-year cycle. There will be two groups of code development committees and they will meet in separate years. The current groups will be reworked. With the energy provisions of the *International Energy Conservation Code*® (IECC®) and Chapter 11 of the *International Residential Code*® (IRC®) now moved to the Code Council's Standards Development Process, the reduced volume of code changes will be distributed between Groups A and B.

Code change proposals submitted for code sections that have a letter designation in front of them will be heard by the respective committee responsible for such code sections. Because different committees hold Committee Action Hearings in different years, proposals for most codes will be heard by committees in both the 2024 (Group A) and the 2025 (Group B) code development cycles. It is very important that anyone submitting code change proposals understands which code development committee is responsible for the section of the code that is the subject of the code change proposal.

Please visit the ICC website at iccsafe.org/products-and-services/i-codes/code-development/current-code-development-cycle for further information on the Code Development Committee responsibilities as it becomes available.

Coordination of the I-Codes

The coordination of technical provisions allows the I-Codes to be used as a complete set of complementary documents. Individual codes can also be used in subsets or as stand-alone documents. Some technical provisions that are relevant to more than one subject area are duplicated in multiple model codes.

Italicized Terms

Words and terms defined in Chapter 2, Definitions, are italicized where they appear in code text and the Chapter 2 definitions apply. Although care has been taken to ensure applicable terms are italicized, there may be instances where a defined term has not been italicized or where a term is italicized but the definition found in Chapter 2 is not applicable. For example, Chapter 2 of the *International Building Code*® (IBC®) contains a definition for "*Listed*" that is applicable to equipment, products and services. The term "listed" is also used in that code to refer to a list of items within the code or within a referenced document. For the latter, the Chapter 2 definition would not be applicable.

Adoption of International Code Council Codes and Standards

The International Code Council maintains a copyright in all of its codes and standards. Maintaining copyright allows the Code Council to fund its mission through sales of books in both print and digital format. The Code Council welcomes incorporation by reference of its codes and standards by jurisdictions that recognize and acknowledge the Code Council's copyright in the codes and standards, and further acknowledge the substantial shared value of the public/private partnership for code development between jurisdictions and the Code Council. By making its codes and standards available for incorporation by reference, the Code Council does not waive its copyright in its codes and standards

The Code Council's codes and standards may only be adopted by incorporation by reference in an ordinance passed by the governing body of the jurisdiction. "Incorporation by reference" means that in the adopting ordinance, the governing body cites only the title, edition, relevant sections or subsections (where applicable), and publishing information of the model code or standard, and the actual text of the model code or standard is not included in the ordinance (see graphic, "Adoption of International Code Council Codes and Standards"). The Code Council does not consent to the reproduction of the text of its codes or standards in any ordinance. If the governing body enacts any changes, only the text of those changes or amendments may be included in the ordinance.

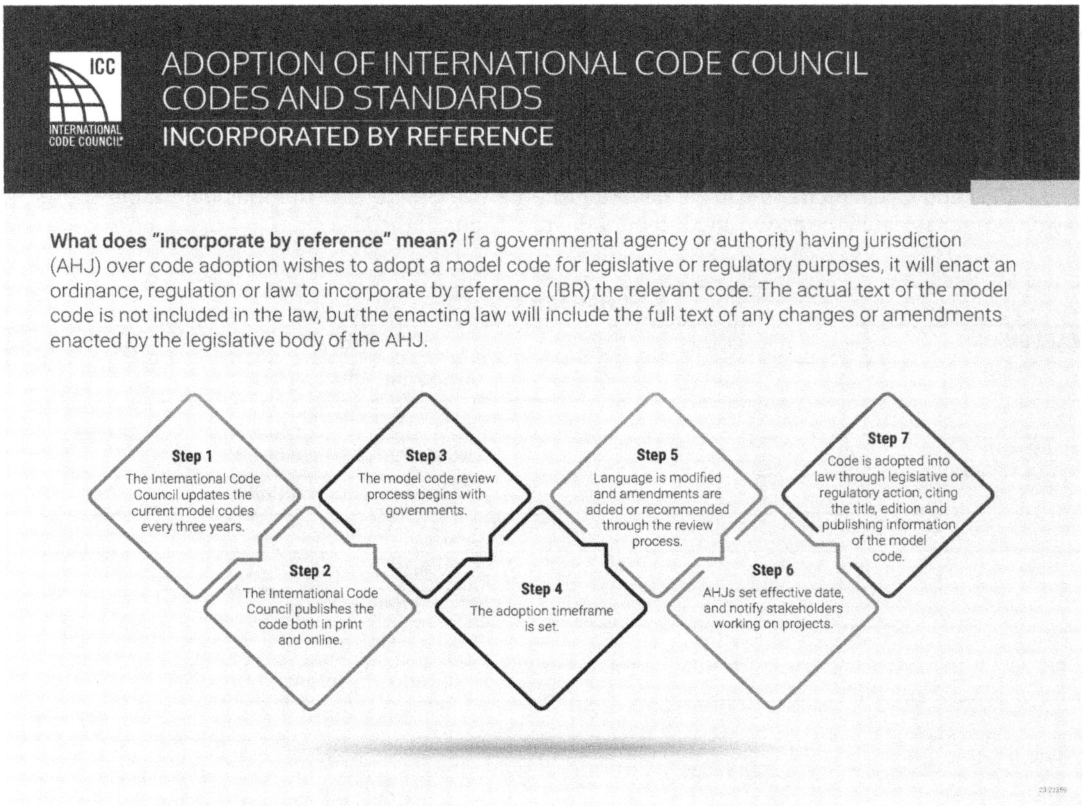

The Code Council also recognizes the need for jurisdictions to make laws accessible to the public. Accordingly, all I-Codes and I-Standards, along with the laws of many jurisdictions, are available to view for free at codes.iccsafe.org/codes/i-codes. These documents may also be purchased, in both digital and print versions, at shop.iccsafe.org.

For further information or assistance with adoption, including a sample ordinance, jurisdictions should contact the Code Council at incorporation@iccsafe.org.

For a list of frequently asked questions (FAQs) addressing a range of foundational topics about the adoption of model codes by jurisdictions and to learn more about the Code Council's code adoption resources, scan the QR code or visit iccsafe.org/code-adoption-resources.

To facilitate adoption of the ICCPC, the jurisdiction must establish the following performance groups for new and/or existing use groups or specific buildings or facilities for the application of this code (see Chapter 3):

ALLOCATION OF USE AND OCCUPANCY CLASSIFICATIONS AND SPECIFIC BUILDINGS OR FACILITIES TO PERFORMANCE GROUPS	
PERFORMANCE GROUP	USE AND OCCUPANCY CLASSIFICATION OR SPECIFIC BUILDINGS OR FACILITIES
I	
II	
III	
IV	

INTRODUCTION TO THE INTERNATIONAL PERFORMANCE CODE

Internationally, the design and regulatory community has embraced the need for a code that emphasizes performance requirements rather than prescriptive requirements. This need is not unique to the international community. As such, the *International Code Council Performance Code® for Buildings and Facilities* (ICCPC®), in this 2024 edition, is designed to meet this need through model code regulations that safeguard the public health and safety in all communities, large and small.

The ICCPC clearly defines the objectives for achieving the intended levels of occupant safety, property protection and community welfare. The code provides a framework to achieve the defined objectives in terms of tolerable levels of damage and magnitudes of design events, such as fire and natural hazards.

The concepts covered by this code are not intended to be any different in scope than those covered by the 2024 edition of the I-Codes. However, this code is distinctly different from the other I-Codes, which, in many cases, direct the user to a single solution to address a safety concern for a building or facility. The ICCPC allows the user to achieve various solutions, systematically. It should be noted that the family of I-Codes provides solutions that will comply with the ICCPC. Conversely, this code provides a procedure to address design and review issues associated with the alternative materials and methods sections of the other I-Codes.

It is strongly recommended that users of this code consult the User's Guide located in the second portion of this publication to gain additional insight into the provisions of this code.

ARRANGEMENT AND FORMAT OF THE 2024 ICCPC

The format of the ICCPC allows each chapter to be devoted to a particular subject. The following table shows how the ICCPC is divided. The chapter synopses detail the scope and intent of the provisions of the ICCPC.

CHAPTER TOPICS	
CHAPTER	**SUBJECTS**
1	Scope and Administration
2	Definitions
3	Design Performance Levels
4	Reliability and Durability
5	Stability
6	Pedestrian Circulation
7	Safety of Users
8	Moisture
9	Interior Environment
10	Mechanical
11	Plumbing
12	Fuel Gas
13	Electricity
14	Energy Efficiency
15	Fire Prevention and Impact Management
16	Management of People
17	Emergency Access and Facilities
18	Emergency Responder Safety
19	Hazardous Materials
20	Referenced Standards
Appendix A	Risk Factors of Use and Occupancy Classifications
Appendix B	Worksheet for Assigning Specific Structures to Performance Groups
Appendix C	Individually Substantiated Design Method
Appendix D	Qualification Characteristics for Design and Review of Performance-Based Designs
Appendix E	Use of Computer Models

Chapter 1 Scope and Administration.

Chapter 1 contains administrative provisions such as intent, scope and requirements related to qualifications, documentation, review, maintenance and change of use or occupancy. Also, provisions for approving acceptable methods are provided.

Chapter 2 Definitions.

Chapter 2 is the repository of the definitions of terms used in the body of the code. The user of the code should be familiar with and consult this chapter because the definitions are essential to the correct interpretation of the code and because the user may not be aware that a term is defined.

Chapter 3 Design Performance Levels.

Chapter 3 sets the framework for determining the appropriate performance desired from a building or facility based on a particular event, such as an earthquake or a fire. Specifically, the user of the code can more easily determine the expected performance level of a building during an earthquake. In the prescriptive codes, the required performance is simply prescribed with no method provided to determine or quantify the level of the building's or facility's performance.

Chapter 4 Reliability and Durability.

Chapter 4 deals with the topics of reliability and durability and how these issues interact with the overall performance of a building or facility over its life. This issue has always been relevant to codes and standards but becomes more obvious when a performance code requires a designer to regard buildings as a system. Reliability includes redundancy, maintenance, durability, quality of installation, integrity of the design and, generally, the qualifications of those involved in this process.

Chapters 5 through 19 (technical provisions).

Chapters 5 through 19 address the technical provisions referred to in the preceding Chapter Topics table and provide topic-specific qualitative statements of intent that relate to current prescriptive code requirements. Generally, such topic-specific qualitative statements are the basic elements missing from prescriptive codes. The statements follow a particular hierarchy, described below.

Objective. The objectives define what is expected in terms of societal goals or what society "demands" from buildings and facilities. Objectives are topic-specific and deal with particular aspects of performance required in a building, such as safeguarding people during escape and rescue.

Functional Statement. The functional statement explains, in general terms, the function that a building must provide to meet the objective or what "supply" must be provided to meet the "demand." For example, a building must be constructed to allow people adequate time to reach a place of safety without exposure to untenable conditions.

Performance Requirement. Performance requirements are detailed statements that break down the functional statements into measurable terms. This is where a link can be more readily made to acceptable methods.

Chapter 20 Referenced Standards.

Chapter 20 lists all of the product and installation standards and codes that are referenced throughout Chapters 1 through 19 and includes identification of the promulgators and the section numbers in which the standards and codes are referenced. These standards and codes become an enforceable part of the code (to the prescribed extent of the reference) as if printed in the body of the code.

Appendices A through E.

Appendices A through E relate to specific provisions of this code and are discussed within the User's Guide as applicable.

GUIDE TO THE USE OF THE INTERNATIONAL CODE COUNCIL PERFORMANCE CODE FOR BUILDINGS AND FACILITIES

The following describes the procedural steps for new and existing buildings.

Procedural Steps for New Buildings.

The following process is an outline for a performance-based design for an entire project or in combination with a prescriptive approach. This procedure for performance-based design extends from design preparation through issuance of a Certificate of Occupancy. The steps are as follows:

1. Preparation of a concept report in accordance with Section 102.3.4.2.1 by a qualified design professional.
2. Design preparation by a design team headed by a qualified principal design professional.
3. Coordination and verification via the principal design professional as a design team leader, with other design professionals, owners and contractors, when applicable.
4. Submit plans and supporting documents to the code official that shall identify which portions of the design are performance based and which portions are based on the prescriptive code. The submittal must include deed restrictions proposed to cover future maintenance requirements and special conditions for the life of the building.
5. Plan review is to be conducted by the code official staff when qualified for performance-based design.
    - 5.1. When staff is deemed not qualified for a proposed project, acquire qualified contract review services.
    - 5.2. Peer review is an optional approach for obtaining an additional review that is supplemental to the plan review.
6. The code official verifies that applicable prescriptive code provisions and performance-based objectives are met. When special inspections are required, ensure that documentation is complete.
7. The code official approves plans and issues a permit.
8. The holder of the permit is responsible to construct in accordance with approved plans and documents.
9. The code official ensures that qualified inspection services are provided and documented where required in accordance with the performance-based code and other applicable codes, and testing requirements are met as follows:
    - 9.1. Phase inspections (reference IBC and other I-Codes).
    - 9.2. Special inspection (reference IBC).
    - 9.3. Testing where required by design documents.
    - 9.4. Documentation that all requirements are met.
10. Issue Certificate of Occupancy with applicable conditions, where required by the approved design documents.

Procedural Steps for Existing Buildings.

For significant remodeling, alterations and additions, the design professional shall:

1. Examine applicable design documents, deed restrictions and maintenance requirements to determine building requirements where the original design is performance based in nature; prepare a concept report in accordance with Section 102.3.4.2.1.
2. Any features based on a performance approach need to be clearly differentiated from features of a building or facility designed using a prescriptive approach.
3. Verify compliance with the operations and maintenance manual.
4. Prepare a report specifying impact and requirements for the proposed design.
5. Prepare design documents based on applicable performance, prescriptive or combination of code provisions and specify which codes are applicable for each portion of the design, including any steps to correct identified deficiencies.
6. Submit reports to the code official for review and acceptance, similar to procedural steps for a new building.

For change of use with no proposed physical alteration, the design professional shall:

1. Document existing building features and systems that impact fire or emergency performance.
2. Verify compliance with the operations and maintenance manual.
3. Prepare appropriate design fire scenarios pertinent to the building or facility and actual use, considering existing mitigation strategies and protection features.
4. Evaluate performance against Section 304, Maximum Level of Impact or Damage to Be Tolerated.
5. Prepare a report detailing impact; design and test systems to the objectives of this code.
6. Submit for review and approval in accordance with Chapter 1.

Flow Chart

The following chart is provided to give guidance as to how the ICCPC is intended to work. Essentially, this chart walks the user through the steps of applying the code. These steps begin with understanding the administrative process and the objectives of the ICCPC and eventually determining the acceptable methods used to design, construct, test, inspect and maintain the building or facility.

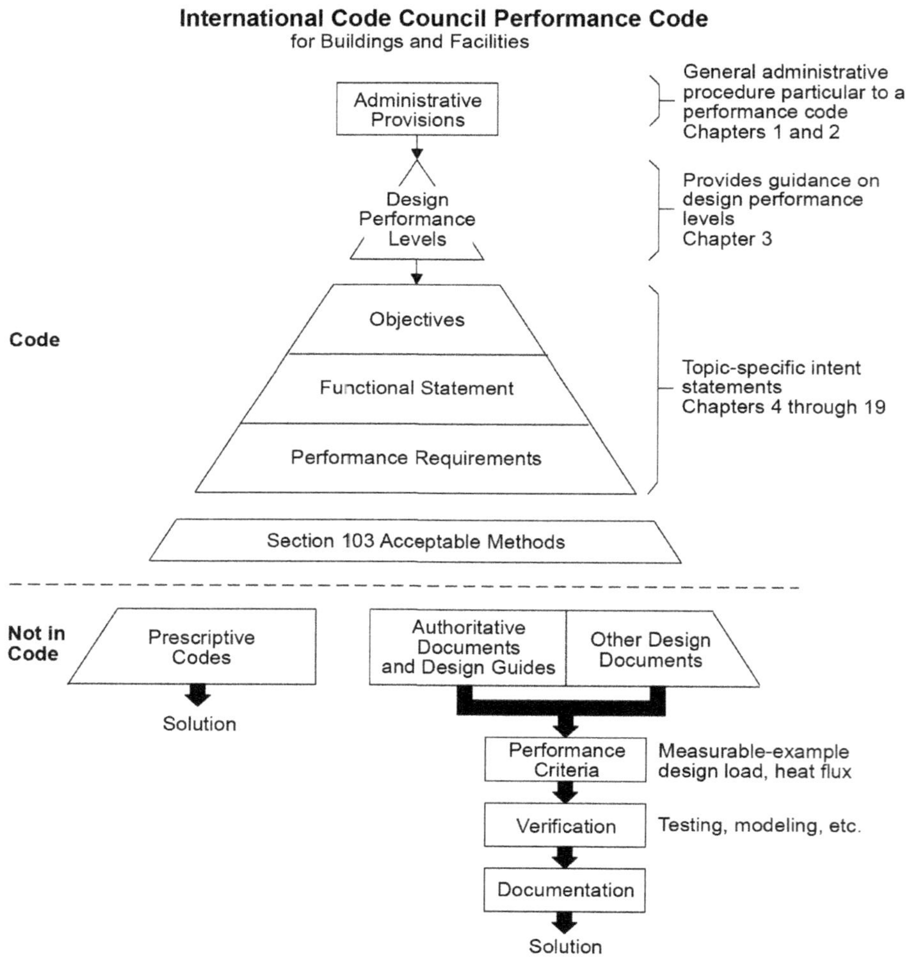

RELOCATION OF CHAPTERS OR SECTIONS

The following table indicates relocation of chapters and sections in the 2024 edition of the ICCPC from the 2021 edition.

RELOCATION	
2024 LOCATION	**2021 LOCATION**
6	7, 19
7	8
8	9
9	10
10	11
11	12
12	13
13	14
14	15
15	Section 601, Section 602, 16, 17
16	18
17	20
18	21
19	22

CONTENTS

PART I—ADMINISTRATIVE

CHAPTER 1 SCOPE AND ADMINISTRATION

User notes:

About this chapter: *Chapter 1 establishes the limits of applicability of the code and describes how the code is to be applied and enforced. Chapter 1 is in two parts: Part 1—Scope and Application (Section 101) and Part 2—Administration and Enforcement (Sections 102 and 103). The scope statements encompass all portions of the code and provide an overall understanding of the limits and applications of the document. The administrative section discusses how this code works in terms of the practical application of the code including stakeholder qualifications and responsibilities, document submittals, and review and construction verification techniques to demonstrate that performance code objectives have been satisfied. Additionally, this section emphasizes the importance of the long-term maintenance needs of a performance-based design and the management of changes to those designs, whether such changes are large or small.*

QR code use: *A QR code is placed at the beginning of any section that has undergone technical revision. To see those revisions, scan the QR code with a smart device or enter the 7-digit code beneath the QR code at the end of the following URL: qr.iccsafe.org/ (see Formatting Changes to the 2024 International Codes for more information).*

PART 1—SCOPE AND APPLICATION

SECTION 101—GENERAL

Scan for Changes e7c1e5d

[A] 101.1 Title. These regulations shall be known as the Performance Code of [NAME OF JURISDICTION], hereinafter referred to as "this code."

[A] 101.2 Purpose. To provide appropriate health, safety, welfare, and social and economic value, while promoting innovative, flexible and responsive solutions that optimize the expenditure and consumption of resources.

[A] 101.3 Scope. This code provides requirements for buildings and structures and includes provisions for structural strength, stability, sanitation, means of access and egress, light and ventilation, safety to life and protection of property from fire and, in general, to secure life and property from other hazards affecting the built environment. This code includes: provisions for the use and occupancy of buildings, structures, facilities and premises; their alteration, repair, maintenance, removal and demolition; the installation and maintenance of amenities including, but not limited to, such services as the electrical, gas, mechanical, plumbing, energy conservation and building transportation systems; and for the storage, handling and use of explosive, flammable and combustible materials, hazardous materials and dangerous operations and processes.

[A] 101.4 Intent. The purpose of this code is to provide an acceptable level of health, safety, and general welfare and to limit damage to property from events that are expected to impact buildings, structures, facilities, equipment and processes. Accordingly, this code intends buildings and structures to provide for the following:

1. An environment free of unreasonable risk of death and injury from fires, explosions or dangerous conditions.
2. A structure that will withstand loads associated with normal use and of the severity associated with the location in which the structure is constructed.
3. Means of egress and access for normal and emergency circumstances.
4. Limited spread of fire both within the building and to adjacent properties.
5. Ventilation and sanitation facilities to maintain the health of the occupants.
6. Natural light, heating, cooking and other amenities necessary for the well-being of the occupants.
7. Efficient use of energy.
8. Safety to firefighters and emergency responders during emergency operations.

PART 2—ADMINISTRATION AND ENFORCEMENT

SECTION 102—ADMINISTRATIVE PROVISIONS

[A] 102.1 Objective. To achieve and maintain the level of safety intended by the code.

Scan for Changes
72f61d7

[A] 102.2 Functional statements.

[A] 102.2.1 Qualifications. *Registered design professionals* shall possess the knowledge, skills and abilities necessary to demonstrate compliance with this code.

[A] 102.2.2 Construction document preparation. *Construction documents* required by this code shall be prepared in adequate detail and submitted for review and approval.

[A] 102.2.3 Review. *Construction documents* submitted in accordance with this code shall be reviewed for code compliance with the appropriate code provisions.

[A] 102.2.4 Construction. Construction shall comply with approved *construction documents* submitted in accordance with this code, and shall be verified and approved to demonstrate compliance with this code.

[A] 102.2.5 Facilities and premises. Facilities and premises shall comply with approved design *construction documents* submitted in accordance with this code, and shall be verified and approved to demonstrate compliance with this code.

[A] 102.2.6 Equipment and processes. Equipment and processes and their installation and operation shall comply with approved *construction documents* submitted in accordance with this code, and shall be verified and approved to demonstrate compliance with this code.

[A] 102.2.7 Materials and contents. Materials and contents shall comply with approved *construction documents* submitted in accordance with this code, and shall be verified and approved to demonstrate compliance with this code.

[A] 102.2.8 Facility operating policies and procedures. Policies, operations, training and procedures shall comply with approved documents submitted in accordance with this code, and shall be verified and approved to demonstrate compliance with this code.

[A] 102.2.9 Supplemental enforcement. Administrative provisions of the International Code Council's family of codes regarding *plan review*, permit issue, inspection and enforcement shall supplement these provisions.

[A] 102.2.10 Maintenance. Maintenance of the *performance-based design* shall be ensured through the issuance and renewal of certificates over the life of the building.

[A] 102.2.11 Management of change. Written procedures managing change to original *construction documents*, system processes, technology, equipment and facilities shall be established and implemented.

[A] 102.2.12 Expected emergency response. *Construction documents* shall clearly describe the level of response expected by emergency responders.

[A] 102.3 Performance requirements.

[A] 102.3.1 Building owner's, or the owner's authorized agent's, responsibility.

[A] 102.3.1.1 Registered design professional. The owner or the owner's authorized agent shall have the responsibility of retaining and furnishing the services of a *registered design professional*, who shall be in responsible charge of preparing and coordinating a complete and comprehensive set of *construction documents* and other services required to prepare reports and other documents in accordance with this code. If the services required by this section are not provided, the use of this code is prohibited.

[A] 102.3.1.2 Registered design professional in responsible charge. Where the project requires the services of multiple *registered design professionals*, a *registered design professional in responsible charge* shall be retained and furnished who shall have the contractual responsibility and authority over all required *registered design professional* disciplines to prepare and coordinate a complete and comprehensive set of *construction documents* for the project.

[A] 102.3.1.3 Peer review. The owner or the owner's authorized agent shall be responsible for retaining and furnishing the services of a *registered design professional* or recognized expert, who will perform as a peer reviewer, where required and approved by the code official. See Section 102.3.6.3 of this code.

[A] 102.3.1.4 Costs. The costs of special services, including *contract review*, where required by the code official, shall be borne by the owner or the owner's authorized agent.

[A] 102.3.1.5 Document retention. The owner or the owner's authorized agent shall retain on the premises documents and reports required by this code and make them available to the code official upon request.

[A] 102.3.1.6 Maintenance. The owner or the owner's authorized agent is responsible to operate and maintain a building, structure or facility designed and built under this code in accordance with the *bounding conditions* and the operations and maintenance manual.

[A] 102.3.1.7 Changes. The owner or the owner's authorized agent shall be responsible to ensure that any change to the facility, process or system does not increase the hazard level beyond that originally designed without approval and that changes shall be documented in accordance with this code.

[A] 102.3.1.8 Special expert. Where the scope of work is limited or focused in an area that does not require the services of a *registered design professional* or the special knowledge and skills associated with the practice of architecture or engineering, a *special expert* may be employed by the owner or the owner's authorized agent as the person in responsible charge of the limited or focused activity. It is the intent of this code that the individual shall possess the qualification characteristics required in Appendix D.

[A] 102.3.1.9 Occupant requirements. The owner or the owner's authorized agent is responsible and accountable to ensure that occupants and employees who are required to take certain actions or perform certain functions in accordance with a *performance-based design* possess the required knowledge and skills and are empowered to perform those actions.

[A] 102.3.2 Registered design professional qualifications. The *registered design professional in responsible charge*, architects, engineers and other *registered design professionals in responsible charge* of their discipline as a member of a design team shall be responsible and accountable to possess the required knowledge and skills to perform design, analysis and verification in accordance with the provisions of this code and applicable professional standards of practice. It is the intent of this code that these individuals possess the qualification characteristics as stated in Appendix D. Qualification statements shall be submitted to the code official for the *registered design professional in responsible charge*, *registered design professionals* and *special experts* to demonstrate compliance with Appendix D.

[A] 102.3.3 Registered design professionals' and special experts' responsibilities.

[A] 102.3.3.1 Registered design professional in responsible charge. Where multiple design disciplines are involved, the *registered design professional in responsible charge* is responsible to ensure that design elements are comprehensive and complete before submittals are made to the code official. During the code review process all designated reports, drawings and *construction documents* necessary to demonstrate compliance with the code shall be submitted by the *registered design professional in responsible charge*. The responsibilities of the *registered design professional in responsible charge* include those of a *registered design professional*.

[A] 102.3.3.2 Responsibilities. *Registered design professionals* are responsible to apply the performance requirements and *acceptable methods* approach in Section 103.3 for performance-based designs where using this code. This code requires design analysis and support documentation to demonstrate the design approach and to verify design objectives and compliance with this code.

[A] 102.3.3.3 Supporting documentation. *Registered design professionals* have the responsibility to provide the appropriate design analysis, research, computations and documentation to demonstrate compliance with applicable performance requirements of this code and applicable *prescriptive code* provisions.

[A] 102.3.3.4 Acceptable methods. *Registered design professionals* shall use *authoritative documents* or *design guides* to determine testing and verification methods for selecting building materials that are compatible with the building systems approach selected.

[A] 102.3.3.5 References. *Registered design professionals* are responsible to document applicable *design guides* or *authoritative documents* for a *performance-based design* and demonstrate how these documents are utilized to substantiate design solutions to show compliance with the provisions of this code. The use of documents that are not accepted as authoritative documents or *design guides* requires substantiation with the code official to obtain acceptance.

[A] 102.3.3.6 Documentation of *bounding conditions*. The *registered design professional* shall document all bounding conditions and establish thresholds that determine when changes must be approved by the code official.

[A] 102.3.3.7 Compliance with *bounding conditions*. The *registered design professional* shall review the completed construction elements, equipment, furnishings, processes and contents to verify compliance with the *bounding conditions* and the critical design features identified in the approved *construction documents*. The code official may require that the *registered design professional in responsible charge* file a report to verify compliance with the bounding conditions and the critical design features at the completion of the project as a condition of obtaining required certificates.

[A] 102.3.3.8 Special expert. The scope of work of a *special expert* shall be limited to the area of expertise as demonstrated in the documentation submitted to the code official for review and approval. Where a *special expert* performs functions of a design, the *special expert* shall assume the responsibilities of that phase of the design.

[A] 102.3.4 Design documentation.

[A] 102.3.4.1 General. The *registered design professional* shall prepare appropriate documentation for the project that clearly provides the design approach and rationale for design submittal, construction and future use of the building, facility or process.

[A] 102.3.4.1.1 Required documentation. The documentation for the project shall identify the goals and objectives; the steps undertaken in the analytical analysis; the facility maintenance and testing requirements; and limitations and restrictions on the use of the facility in order to stay within *bounding conditions*. Where requirements for documentation are specified in applicable engineering or *design guides*, documentation shall be included in the *construction documents*. Computer modeling documentation shall comply with Appendix E.

[A] 102.3.4.1.2 Extent of documentation. The level of documentation provided shall be adequate to convey the required information clearly to the involved parties and shall be commensurate with the scope and complexity of the project.

[A] 102.3.4.1.3 Verification of compliance. Documentation shall be prepared that clearly verifies that applicable performance and applicable *prescriptive code* provisions have been met.

[A] 102.3.4.1.4 Deed restriction. Design features with *bounding conditions* that require continued maintenance or supervision by the owner or the owner's authorized agent throughout the life of the building, facility or process as conditions of compliance with the objectives of this code shall be recorded as a deed restriction until released by the code official. Where required by the code official, the deed restriction shall be modified to reflect specific changes.

[A] 102.3.4.1.5 Phased and partial occupancy. The *construction documents* shall include an evaluation of hazards and proposed resolution of associated risks during construction in advance of a request for phased or partial occupancy.

102.3.4.1.6 Emergency response capabilities. Design documentation shall clearly describe the level of response expected by emergency responders under the direct control of the owner or the owner's authorized agent. Emergency response capabilities, staffing levels, training requirements and equipment availability shall be documented as a *bounding condition*.

[A] 102.3.4.2 Reports and manuals. Where required by the code official, design documentation shall include a concept report, design report and operations and maintenance manual.

[A] 102.3.4.2.1 Concept report. The concept report shall document the preliminary details of the project, identify the parties involved in the project, and define the goals and objectives to be utilized in the performance-based design analysis. The concept report shall be submitted to the code official as a means of communicating the programming and early schematic phase of a proposed project and to obtain concurrence between the code official and the project design team on the goals and objectives to be utilized in the analysis. The concept report shall address but not be limited to the following:

1. General project information, including schematic layout and site plan.
2. Definition of project scope.
3. Description of building and occupant characteristics.
4. Project goals and objectives.
5. Selected event scenarios.
6. Methods of evaluation.
7. Qualification statements for the *registered design professional in responsible charge*, *registered design professionals*, *peer reviewers* and *special experts*.
8. Proposed performance and *prescriptive code* usage.
9. Conceptual site and building plan.

[A] 102.3.4.2.2 Design report. The design report shall document the steps taken in the design analysis, clearly identifying the criteria, parameters, inputs, assumptions, sensitivities and limitations involved in the analysis. The design report shall clearly identify *bounding conditions*, assumptions and sensitivities that clarify the expected uses and limitations of the performance analysis. This report shall verify that the design approach is in compliance with the applicable codes and *acceptable methods* and shall be submitted for concurrence by the code official prior to the *construction documents* being completed. The report shall document the design features to be incorporated based on the analysis. The design report shall address but not be limited to the following:

1. Project scope.
2. Goals and objectives.
3. Performance criteria.
4. Hazard scenarios.
5. Design fire loads and hazards.
6. Final design.
7. Evaluation and peer review.
8. *Bounding conditions* and critical design assumptions.
9. Critical design features.
10. System design and operational requirements.
11. Operational and maintenance requirements.
12. *Commissioning* testing requirements and acceptance criteria.
13. Frequency of certificate renewal.
14. Supporting documents and references.
15. Preliminary site and floor plans.

[A] 102.3.4.2.3 Operations and maintenance manual. The operations and maintenance manual shall identify system and component *commissioning* requirements and the required interactions between these systems. The manual shall identify for the facility owner or the owner's authorized agent and the facility operator those actions that need to be performed on a regular basis to ensure that the components of the *performance-based design* are in place and operating properly. Furthermore, the operations and maintenance manual shall identify the restrictions or limitations placed on the use and operation of the facility in order to stay within the *bounding conditions* of the *performance-based design*. The operations and maintenance manual shall be submitted at the time of the *construction documents* submittal, unless the code official approves another time based on the type of project and data needed for a composite review. The operations and maintenance manual shall address but not be limited to the following:

1. Description of critical systems.
2. Description of required system interactions.
3. Occupant responsibilities.
4. Occupant and staff training requirements.
5. Periodic operational requirements.
6. Periodic maintenance requirements.
7. Periodic testing requirements.
8. Limitations on facility operations (due to *bounding conditions*).
9. Report format for recording maintenance and operation data.
10. System and component *commissioning* requirements.

[A] 102.3.5 Design submittal.

[A] 102.3.5.1 General. Applicable *construction documents* required in Sections 102.3.2, 102.3.3 and 102.3.4 for submittal in this code and other applicable codes under the jurisdiction of the code official shall be submitted to the code official for review. The documents shall be submitted in accordance with the jurisdiction's procedures and in sufficient detail to obtain appropriate permits.

[A] 102.3.5.2 Coordination of construction documents. Design documents shall be coordinated by the *registered design professional in responsible charge* for consistency, compatibility and completeness prior to submittal. Documentation shall be provided to the code official to demonstrate compliance with the performance provisions, including *acceptable methods*.

[A] 102.3.5.3 Performance-based design features. The *construction documents* shall clearly indicate those areas of the design that are performance-based and shall be provided to the code official.

[A] 102.3.5.4 Extent of documentation and references. The code official shall be provided with sufficient documentation to support the validity, accuracy, relevance and precision of the proposed methods. Copies of referenced documentation shall be made available to the code official.

[A] 102.3.5.5 Inspections, testing, operation and maintenance. The *construction documents* shall specify when and where special inspection and testing are required, the standards of acceptance for demonstrating compliance with the *construction documents*, and operations and maintenance requirements for future use of the building.

[A] 102.3.5.6 Management of change. The submittal shall include appropriate management of change protocol to address how changes in the *construction documents* will be managed for construction, operation and maintenance activities.

[A] 102.3.6 Review and approval.

[A] 102.3.6.1 Procedures. Document review and approval shall be accomplished in accordance with the code official's procedures.

[A] 102.3.6.2 Review. The code official shall be responsible to perform a knowledgeable review of the proposed design project to verify compliance with this code, or the code official shall retain competent assistance to perform the review in accordance with acceptable standards of practice.

[A] 102.3.6.3 Contract and peer review. Review may be accomplished by a contract reviewer where the reviewer is assigned by the code official. In addition, the code official may require a peer review process to review design criteria and supporting documents and *construction documents*.

[A] 102.3.6.4 Approval. After documents and other supporting data are reviewed and approved by the code official to verify compliance with the applicable codes, permits may be issued.

[A] 102.3.7 Permits and inspections.

[A] 102.3.7.1 Permits. Prior to the start of construction, appropriate permits shall be obtained in accordance with the jurisdiction's procedures and applicable codes.

[A] 102.3.7.2 Inspection. Approved inspections shall be obtained in accordance with the *construction documents*, jurisdiction's procedures and applicable codes.

[A] 102.3.7.3 Verification reports. Inspection, testing and related verification reports shall be filed with the code official to verify compliance with approved *construction documents* and applicable *prescriptive code* provisions.

[A] 102.3.7.4 Product installation. Compliance shall be verified for materials, fabrication, manufacturer's and engineer's installation procedures by product labeling, certification, *quality assurance* processes and testing, as applicable, to verify compliance.

[A] 102.3.7.5 Compliance verification. At the completion of construction, the code official shall verify that inspection and testing reports demonstrate compliance with the applicable codes and approved *construction documents*.

[A] 102.3.7.6 Operational permits. Prior to initiating facility uses and processes regulated under Part III of this code, appropriate permits shall be obtained.

[A] 102.3.8 Project documentation.

[A] 102.3.8.1 Verification of compliance. Upon completion of the project, documentation shall be prepared that verifies performance and *prescriptive code* provisions have been met. Where required by the code official in accordance with Section 102.3.3.6, the *registered design professional in responsible charge* shall file a report that verifies *bounding conditions* are met.

[A] 102.3.8.2 Extent of documentation. Approved *construction documents*, the operations and maintenance manual, inspection and testing records, and certificates of occupancy with conditions shall be included in the project documentation of the code official's records.

[A] 102.3.8.3 Deed restrictions. Design features with *bounding conditions* determined by the *registered design professional* to require continued operation and maintenance by the owner or the owner's authorized agent throughout the life of the building as conditions of compliance with the objectives of this code shall be recorded as a deed restriction as required by the code official until released by the code official.

[A] 102.3.8.4 Technical opinion. The code official has the authority to require a technical opinion and report from an individual or organization with special expertise to identify and develop methods of protection from special hazards and to determine the acceptability of technologies, processes, products, equipment, materials and uses applicable to the design, operation or use of a building or facility. The intent of this code is that the technical opinion and report shall be prepared by a qualified individual. See Appendix D.

[A] 102.3.9 Certificates.

[A] 102.3.9.1 Certificate of occupancy. Prior to occupancy of a building, a certificate of occupancy shall be obtained from the code official.

[A] 102.3.9.1.1 Continued occupancy. A certificate of occupancy is required for the continued occupancy of a building.

[A] 102.3.9.1.2 Temporary certificate of occupancy. The code official has the authority to issue a temporary certificate of occupancy for a limited time with specified conditions, providing all life-safety items are accepted.

[A] 102.3.9.1.3 Conditional certificate of occupancy. The code official has the authority to issue a certificate of occupancy with conditions valid for a specified time period that requires continued compliance with *bounding conditions* and the operations and maintenance manual. Failure to maintain compliance with the conditions of the certificate of occupancy is a violation of this code.

[A] 102.3.9.1.4 Revocation and renewal. Failure of the building owner or the owner's authorized agent to demonstrate to the code official that the building is being operated and maintained in compliance with Sections 102.3.1.6 and 102.3.9.1 is cause to revoke or not renew a certificate of occupancy.

[A] 102.3.9.2 Certificate of compliance. Prior to use of a building, facility, process or premises subject to Part III of this code, a certificate of compliance shall be obtained from the code official.

[A] 102.3.9.2.1 Continued use. A certificate of compliance is required for the continued use or occupancy of a facility, process or equipment subject to Part III of this code throughout the life of the facility.

[A] 102.3.9.2.2 Renewal frequency. The certificate of compliance issued subject to Part III of this code shall be renewed at a frequency as determined in the design and approved by the code official.

[A] 102.3.9.2.3 Revocation and renewal. Failure of the owner or the owner's authorized agent to demonstrate compliance with this section is cause to revoke or not renew the certificate of compliance.

[A] 102.3.10 Maintenance.

[A] 102.3.10.1 Owner, or the owner's authorized agent's, responsibility. The owner or the owner's authorized agent is responsible for maintaining the building or facility in accordance with the approved documents.

[A] 102.3.10.2 Continued compliance. Compliance with the operations and maintenance manual and bounding conditions shall be verified throughout the life of the building or facility at a frequency in accordance with the approved documents.

[A] 102.3.10.3 Compliance verification. Documents verifying that the building, facilities, premises, processes and contents are in compliance with the approved *construction documents* and are maintained in a safe manner shall be filed with the code official at a frequency approved by the code official.

[A] 102.3.11 Remodeling, addition or change/approval of use.

[A] 102.3.11.1 Analysis of change. The *registered design professional* shall evaluate the existing building, facilities, premises, processes, contents and the applicable documentation of the proposed change as it affects portions of the building, facility, premises, processes and contents that were previously designed for compliance under a performance-based code. Prior to any change that was not documented in a previously approved design, the *registered design professional in responsible charge* shall examine the applicable design documents, *bounding conditions*, operation and maintenance manuals, and deed restrictions.

[A] 102.3.11.2 Coordination of design. Where multiple design disciplines are involved, one *registered design professional* shall be responsible to ensure that design elements are comprehensive and complete before submittals are made to the code official. During the code review process, designated reports, drawings and *construction documents* necessary to demonstrate compliance with the code shall be submitted by the *registered design professional*.

[A] 102.3.11.3 Change in activity or contents. Any change in activity or contents that results in an increase in hazard or risk that exceeds the *bounding conditions* requires an evaluation and approval. The code official shall have the authority to require a full evaluation of the design.

[A] 102.3.11.4 Additions, renovations and related construction changes. Construction activities in existing buildings, facilities, premises or processes shall be evaluated by a *registered design professional* and documented in a written report, which shall be submitted for review and approval in conjunction with the permit request. The report shall identify whether the proposed construction exceeds the *bounding conditions*, which will result in an increase in hazard or risk beyond that expected in the approved original *construction document*. Where *bounding conditions* are not exceeded, the original construction document need not be revised. Where *bounding conditions* are exceeded, the original *construction document* shall be revised so that compliance with this code is perpetuated.

[A] 102.3.11.5 Designs exceeding *bounding conditions*. Where a proposed change exceeds the bounding conditions and does not result in an increase to hazard or risk, as approved by the code official, any person authorized by the laws of the jurisdiction is allowed to prepare *construction documents* and reports for submittal.

[A] 102.3.11.6 Change in design objectives and *bounding conditions*. Where changes are proposed to the design objectives and bounding conditions of an existing building, facility, process or contents, a written report by the *registered design professional* shall be prepared to specify the new design objectives and demonstrate compliance with the current code.

[A] 102.3.12 Administration and enforcement.

[A] 102.3.12.1 Supplemental administrative provisions. Administrative provisions of the International Code Council's family of codes shall supplement the performance provisions for *plan review*, permit issuance, inspection, certificate of occupancy or compliance, and enforcement.

[A] 102.3.13 Violations.

[A] 102.3.13.1 Unlawful acts. It shall be unlawful for any person, firm or corporation to erect, construct, alter, extend, repair, move, remove, demolish or occupy any building, structure or facility regulated by this code, or cause same to be done, in conflict with or in violation of any of the provisions of this code.

[A] 102.3.13.2 Notice of violation. The code official shall serve a notice of violation or order on the person responsible for the erection, construction, alteration, extension, repair, moving, removal, demolition or occupancy of a building or facility in violation of the provisions of this code or in violation of a detail statement or *construction documents* approved thereunder, or in violation of a permit or certificate issued under the provisions of this code. Such order shall direct the discontinuance of the illegal action or condition and the abatement of the violation.

[A] 102.3.13.3 Violation. If the notice of violation is not complied with promptly, the code official has the authority to request the legal counsel of the jurisdiction to institute the appropriate proceeding at law or in equity to restrain, correct or abate such violation, or to require the removal or termination of the unlawful occupancy of the building or structure in violation of the provisions of this code or of the order or direction made pursuant thereto.

[A] 102.3.13.4 Penalties. Any person who violates a provision of this code or fails to comply with any of the requirements thereof or who erects, constructs, alters or repairs a building, structure or facility in violation of the approved *construction documents* or directive of the code official or of a permit or certificate issued under the provisions of this code shall be subject to penalties as prescribed by law.

SECTION 103—ACCEPTABLE METHODS

[A] 103.1 Objective. To require the use of recognized *authoritative documents* or *design guides* for analysis, measurement of performance and determination of criteria used to evaluate compliance with the performance requirements of this code. See Chapter 2 for definitions.

[A] 103.2 Functional statements.

[A] 103.2.1 Approved methodologies. Design approaches shall utilize *authoritative documents* and *design guides* to demonstrate that designs are based on applicable and valid technical and scientific methodologies.

[A] 103.2.2 Construction documents. *Construction documents* shall indicate the method by which the design and construction are to be verified and applicable systems are to be measured.

[A] 103.2.3 Testing and inspection. Testing and inspection of materials and systems shall be based on applicable *authoritative documents* and *design guides*.

[A] 103.3 Performance requirements and acceptance method approach.

[A] 103.3.1 Construction documents. *Registered design professionals* shall utilize *acceptable methods*. *Construction documents* shall contain the design approach, analysis, research, computation and criteria for acceptance that specify the applicable *design guides*, and *authoritative documents* utilized to demonstrate that design objectives are met.

[A] 103.3.2 Construction documents. *Construction documents* shall include design verification methods that are required to demonstrate compliance with design objectives and applicable *authoritative documents* and *design guides*.

[A] 103.3.3 Individually substantiated design methods. Documents that do not meet the criteria for *authoritative documents* or *design guides* shall comply with the individually substantiated design method criteria in Appendix C.

[A] 103.3.4 Peer review. Designs that propose to use documents that do not meet the criteria for *authoritative documents* or *design guides* shall not be permitted unless approval is given by the code official. The resulting *performance-based design* shall undergo an independent *peer review* process.

CHAPTER

2 DEFINITIONS

User notes:

About this chapter: *Codes, by their very nature, are technical documents. Every word, term and punctuation mark can add to or change the meaning of a technical requirement. It is necessary to maintain a consensus on the specific meaning of each term contained in the code. Chapter 2 performs this function by stating clearly what specific terms mean for the purpose of the code.*

SECTION 201—GENERAL

[BG] 201.1 Scope. Unless otherwise expressly stated, the following words and terms shall, for the purposes of this code, have the meanings indicated in this chapter.

[BG] 201.2 Interchangeability. Words used in the present tense include the future; words in the masculine gender include the feminine and neuter; the singular number includes the plural and the plural, the singular.

[BG] 201.3 Terms not defined in other codes. Where terms are not defined through the methods authorized by this section, such terms shall have ordinarily accepted meanings such as the context implies.

SECTION 202—DEFINED TERMS

Scan for Changes 3e38867

[A] ACCEPTABLE METHODS. Design, analysis and testing methods that have been approved for use in developing design solutions for compliance with the requirements of this code. See Section 103.

[M] ACCESS (TO). That which enables a device, appliance or equipment to be reached by ready access or by a means that first requires the removal or movement of a panel or similar obstruction [see also "Ready access (to)"].

[BG] AMENITY. An attribute of, or system in, the building that provides services or functions related to the use of the building by the occupants or that contributes to the comfort of the occupants, and that is not necessary for the minimum protection of the occupants. For example, an automatic sprinkler system is not a building amenity.

[A] ARCHITECT/ENGINEER. The individual architect or engineer who is registered or licensed to practice his or her respective design profession as defined by the statutory requirements of the professional registration laws of the state or jurisdiction in which the project is to be constructed. See Qualification Characteristics in Appendix D.

[A] AUTHORITATIVE DOCUMENT. A document containing a body of knowledge commonly used by practicing architects or engineers. It represents the state of the art, including accepted engineering practices, test methods, criteria, loads, safety factors, reliability factors and similar technical matters. The document portrays the standard of care normally observed with a particular discipline. The content is promulgated through an open consensus process or a review by professional peers conducted by recognized authoritative professional societies, codes or standards organizations, or governmental bodies.

[A] BOUNDING CONDITIONS. Conditions that, if exceeded, invalidate the *performance-based design*. These could be maximum allowable conditions such as fuel load or type and arrangement of fuel load that must be maintained throughout the life of a building to ensure that design parameters are not exceeded.

[A] CODE. The term used in this document to refer to the *International Code Council Performance Code for Buildings and Facilities*. Other codes in the International Code Council's family of codes and the *National Electrical Code* are identified where used.

[A] COMMISSIONING. The process of verifying that a system meets design, technical standards and code expectations via inspection, testing and operational functionality.

[A] CONSTRUCTION DOCUMENTS. Design drawings and written, graphic and pictorial documents prepared or assembled for describing the design, location and physical characteristics of the elements of a project necessary for obtaining a permit.

[A] CONSULTANT. An individual who provides specialized services to an owner, designer, code official or contractor.

[A] CONTRACT REVIEW. *Plan review*, as defined below, performed by a *consultant* who is retained by the code official for that purpose.

[A] DESIGN GUIDE. A document containing a body of knowledge or information used by practicing architects and engineers that is not required to meet an open consensus requirement. It represents accepted architectural/engineering principles and practices, tests and test data, criteria, loads, safety factors, reliability factors and similar technical data.

[BS] ESSENTIAL FACILITIES. Buildings and other structures that are intended to remain operational in the event of extreme environmental loading from flood, wind, snow or earthquake.

[A] FACILITY. (General Application) Includes all buildings or structures (permanent or temporary), including all fire- and life-safety systems installed therein. A facility includes interior and exterior storage areas, equipment and processes dealing with flammable and combustible substances and hazardous materials, on site. The term includes tents, membrane structures, mobile and manufactured structures, storage tanks, piers, wharves and all required access roads and areas.

FACILITY. (Only applicable to Section 1502). The entire building or any portion of a building, structure or area, including the site on which such building, structure or area is located, wherein specific services are provided or activities are performed.

[A] PEER REVIEW. An independent and objective technical review of the design of a building or structure to examine the proposed conceptual and analytical concepts, objectives and criteria of the design and construction. It shall be conducted by an *architect* or *engineer* who has a level of experience in the design of projects similar to the one being reviewed at least comparable to that of the *architect* or *engineer* responsible for the project.

[A] PERFORMANCE-BASED DESIGN. An engineering approach to design elements of a building based on agreed upon performance goals and objectives, engineering analysis and quantitative assessment of alternatives against the design goals and objectives using accepted engineering tools, methodologies and performance criteria.

[A] PLAN REVIEW. A review of the *construction documents* by the code official to verify conformance to applicable performance and *prescriptive code* requirements.

[A] PRESCRIPTIVE CODES. The International Code Council's family of codes, which provide specific (design, construction and maintenance) requirements for building, energy conservation, fire prevention, mechanical, plumbing and so forth.

[A] QUALITY ASSURANCE. Inspection by code officials, and special inspection and testing by qualified persons and observation by architects/engineers, where applicable, of the construction of a building or structure to verify general conformance with the *construction documents*, and applicable performance and *prescriptive code* requirements.

[M] READY ACCESS (TO). That which enables a device, appliance or equipment to be directly reached, without requiring the removal or movement of any panel or similar obstruction [see "Access (to)"].

[A] REGISTERED DESIGN PROFESSIONAL. An individual who is registered or licensed to practice his or her respective design profession as defined by the statutory requirements of the professional registration laws of the state or jurisdiction in which the project is to be constructed.

[A] REGISTERED DESIGN PROFESSIONAL IN RESPONSIBLE CHARGE. A *registered design professional* engaged by the owner or the owner's authorized agent to review and coordinate certain aspects of the project, as determined by the code official, for compatibility with the design of the building or structure, including submittal documents prepared by others, deferred submittal documents and phased submittal documents.

[BG] SAFE PLACE. An interior or exterior area wherein protection from hazards is provided by construction or appropriate separation distance.

[BF] SAFETY SYSTEMS. Designed systems in the building provided to serve as the protection for the occupants and the building and contents from hazards.

[BG] SERIOUS INJURY. An injury requiring hospitalization or multiple visits to a healthcare provider to effect treatment.

[A] SPECIAL EXPERT. An individual who has demonstrated qualifications in a specific area, outside the practice of architecture or engineering, by education, training and experience.

[A] THIRD-PARTY REVIEW. A term associated with *quality assurance* and independence from another party whose work product is being reviewed. Third-party review does not apply to the *peer review* process.

CHAPTER

3 DESIGN PERFORMANCE LEVELS

User notes:

About this chapter: *Chapter 3 is unique to this code. It is intended to provide a framework to establish minimum levels to which buildings or facilities should perform when subjected to events such as fires and natural hazards. The minimums established by this chapter are based on the types of risks associated with the use of the building or facility, the intended function of the building or facility and the importance of the building or facility to a community. This information is then compared with the type and sizes of events that may affect the building or facility. As noted in the Effective Use portion of this document, it is intended that this chapter provide a link between the policy makers and the designers. In many respects, this chapter is the performance code equivalent of the height and area requirements, occupancy classifications and related requirements.*

SECTION 301—MINIMUM PERFORMANCE

Scan for Changes 69621f5

[BG] 301.1 Purpose. This chapter provides the basis for developing the acceptable level of design based on building use, risk factors, magnitudes of event and acceptable level of damage. Magnitudes are defined in subsequent chapters of this code but interrelate with this chapter in the development of design methods for the mitigation of hazards.

[BG] 301.2 Objective. To establish performance groups for buildings and other structures and to establish minimum acceptable losses based on those performance groups.

[BG] 301.3 Functional statements.

[BG] 301.3.1 Performance level. The performance of a building or facility is based on the ability of the building or facility to tolerate specified magnitudes of event within tolerable limits of damage.

[BG] 301.3.2 Demonstration of performance. Performance is acceptable where the design performance levels are demonstrated to be met or exceeded, to the satisfaction of the code official, in accordance with the assigned or designated use groups, performance groups, magnitudes of event and maximum tolerable damage limits; and the objectives, functional statements and performance requirements of this code.

SECTION 302—USE AND OCCUPANCY CLASSIFICATION

Scan for Changes a7ed3d9

[BG] 302.1 General. The objective of the assignment of use and occupancy classification is to identify the primary uses of buildings and other structures, and portions of buildings and other structures, and to identify risk factors associated with these uses, in order to facilitate design and construction in accordance with other provisions of this code.

[BG] 302.2 Determination of use. In determining the primary use of a building or other structure, or portion of a building or other structure, the following shall be considered:

1. **Principal purpose or function.** The principal purpose or function of the building or other structure.
2. **Hazards.** The hazard-related risk(s) to the users of the building or other structure.

[BG] 302.3 Guidance. The use and occupancy classifications found in the *International Building Code* shall be permitted to be used for guidance in determining the principal purposes or functions for buildings or other structures.

[BG] 302.4 Risk factors. In determining the hazard-related risk(s) to users of buildings and other structures, the following risk factors shall be considered:

[BG] 302.4.1 Nature of the hazard. The nature of the hazard, whether it is likely to originate internal or external to the building or facility, and how it may impact the occupants, the building or other structure, and the contents.

[BG] 302.4.2 Number of occupants. The number of persons normally occupying, visiting, employed in or otherwise using the building, other structure or portion of the building or other structure.

[BG] 302.4.3 Length of occupancy. The length of time the building or other structure is normally occupied by people.

[BG] 302.4.4 Sleeping characteristics. Whether people normally sleep in the building.

[BG] 302.4.5 Familiarity. Whether the building or other structure occupants and other users are expected to be familiar with the building or other structure layout and means of egress.

[BG] 302.4.6 Vulnerability. Whether a significant percentage of the building or other structure occupants are, or are expected to be, members of vulnerable population groups such as infants, young children, elderly persons, persons with physical disabilities, persons with mental disabilities, or persons with other conditions or impairments that could affect their ability to make decisions, egress without the physical assistance of others or tolerate adverse conditions.

[BG] 302.4.7 Relationships. Whether a significant percentage of building or other structure occupants and other users have family or dependent relationships.

SECTION 303—PERFORMANCE GROUPS

[BG] 303.1 Performance group allocation. Use groups and hazard-related occupancies have been allocated to performance groups using the risk factors identified in Section 302.4. Specific buildings and other structures have been allocated to performance groups using the risk factors identified in Section 302.4 combined with the relative importance of protecting the building or other structure to the community. The allocated performance group shall not be lower than the corresponding risk category determined in accordance with Table 303.1.

Scan for Changes
986ef35

[BS] TABLE 303.1—RISK CATEGORY OF BUILDINGS AND OTHER STRUCTURES

RISK CATEGORY	NATURE OF OCCUPANCY
I	Buildings and other structures that represent a low hazard to human life in the event of failure, including but not limited to: Agricultural facilities. Certain temporary facilities. Minor storage facilities.
II	Buildings and other structures except those listed in Risk Categories I, III and IV.
III	Buildings and other structures that represent a substantial hazard to human life in the event of failure, including but not limited to: Buildings and other structures whose primary occupancy is public assembly with an occupant load greater than 300. Buildings and other structures containing one or more public assembly spaces, each having an occupant load greater than 300 and a cumulative occupant load of the public assembly spaces of greater than 2,500. Buildings and other structures containing Group E or Group I-4 occupancies or a combination thereof, with an occupant load greater than 250. Buildings and other structures containing educational occupancies for students above the 12th grade with an occupant load greater than 500. Group I-3 occupancies. Any other occupancy with an occupant load greater than 5,000.[a] Power-generating stations, water treatment facilities for potable water, wastewater treatment facilities and other public utility facilities not included in Risk Category IV. Buildings and other structures not included in Risk Category IV containing quantities of toxic or explosive materials that: • Exceed maximum allowable quantities per control area as given in Table 307.1(1) or 307.1(2) of the *International Building Code* or per outdoor control area in accordance with the *International Fire Code*. • Are sufficient to pose a threat to the public if released.[b]
IV	Buildings and other structures designated as *essential facilities*, including but not limited to: Group I-2, Condition 2 occupancies. Ambulatory care facilities having emergency surgery or emergency treatment facilities. Fire, rescue, ambulance and police stations and emergency vehicle garages. Designated earthquake, hurricane or other emergency shelters. Designated emergency preparedness, communications and operations centers and other facilities required for emergency response. Power-generating stations and other public utility facilities required as emergency backup facilities for Risk Category IV structures. Buildings and other structures containing quantities of highly toxic materials that: • Exceed maximum allowable quantities per control area as given in Table 307.1(2) of the *International Building Code* or per outdoor control area in accordance with the *International Fire Code*. • Are sufficient to pose a threat to the public if released.[b] Aviation control towers, air traffic control centers and emergency aircraft hangars. Buildings and other structures having critical national defense functions. Water storage facilities and pump structures required to maintain water pressure for fire suppression.

a. For purposes of occupant load calculation, occupancies required by Table 1004.5 of the *International Building Code* to use gross floor area calculations shall be permitted to use net floor areas to determine the total occupant load. The floor area for vehicular drive aisles shall be permitted to be excluded in the determination of net floor area in parking garages.

b. Where approved by the building official, the classification of buildings and other structures as Risk Category III or IV based on their quantities of toxic, highly toxic or explosive materials is permitted to be reduced to Risk Category II, provided that it can be demonstrated by a hazard assessment in accordance with Section 1.5.3 of ASCE 7 that a release of the toxic, highly toxic or explosive materials is not sufficient to pose a threat to the public.

[BG] 303.2 Unique performance group allocation. Where necessary or desired, allocation of specific buildings or other structures to performance groups differing from the corresponding risk category in Table 303.1 is permitted based on the needs specific to a community or owner or if there are unusual circumstances associated with the building or other structures.

[BG] 303.3 Magnitudes of event and level of damage. Performance groups identify the minimum required performance of buildings or other structures through a relationship of the magnitude of an event to the maximum level of impact or damage to be tolerated shown in Table 303.3. The use of Table 303.3 shall be an iterative process. It shall be used to determine the acceptable impact of certain events based on their magnitude, and then used iteratively to evaluate various designed mitigation features. The use of Table 303.3 shall consider, explicitly or implicitly, all four design event magnitudes for the assigned performance group. Assignment of performance groups is accomplished through consideration of building or other structure uses, building or other structure risk factors, and the importance of a building or other structure to a community.

[BG]TABLE 303.3—MAXIMUM LEVEL OF DAMAGE TO BE TOLERATED BASED ON PERFORMANCE GROUPS AND DESIGN EVENT MAGNITUDES

MAGNITUDE OF DESIGN EVENT (INCREASING MAGNITUDE OF EVENT ↑)	INCREASING LEVEL OF PERFORMANCE → PERFORMANCE GROUPS			
	Performance Group I	Performance Group II	Performance Group III	Performance Group IV
VERY LARGE (Very Rare)	SEVERE	SEVERE	HIGH	MODERATE
LARGE (Rare)	SEVERE	HIGH	MODERATE	MILD
MEDIUM (Less Frequent)	HIGH	MODERATE	MILD	MILD
SMALL (Frequent)	MODERATE	MILD	MILD	MILD

[BG] 303.4 Performance groups. There are four performance groups (PG), identified as I, II, III and IV.

[BG] 303.4.1 Performance Group I. The minimum design performance level with which all buildings or other structures shall comply where allocated to Risk Category I in accordance with Table 303.1 or posing a low risk to human life, should the buildings or other structures fail.

[BG] 303.4.2 Performance Group II. The minimum design performance level with which all buildings or other structures subject to this code, except those classified as PG I, PG III or PG IV, shall comply.

[BG] 303.4.3 Performance Group III. The minimum design performance level with which buildings or other structures shall comply where allocated to Risk Category III in accordance with Table 303.1 or of an increased level of societal benefit or importance.

[BG] 303.4.4 Performance Group IV. The minimum design performance level with which buildings or other structures shall comply where allocated to Risk Category IV in accordance with Table 303.1 or that present an unusually high risk or that are deemed essential.

[BG] 303.5 Alternative performance group designations. The performance group for specific buildings or other structures or classes of buildings or other structures is permitted to be redesignated with the approval of the code official. If a higher design performance level is desired, the design team, with the approval of the code official, shall be permitted to choose a higher performance group. For existing buildings or other structures, the code official is authorized to adjust tolerable limits of impact to a building or other structure and its contents.

SECTION 304—MAXIMUM LEVEL OF IMPACT OR DAMAGE TO BE TOLERATED

Scan for Changes
7460978

[BG] 304.1 General. Design performance levels establish how a building or other structure is expected to perform, in terms of tolerable limits, under varying load conditions. For each magnitude of event (small to very large), considered as a design load, based on realistic event scenarios, the design shall provide high confidence that the corresponding maximum level of damage to be tolerated for the appropriate performance group will be met. This relationship is illustrated in Table 303.3.

[BG] 304.2 Level of impact or damage. There are four design performance levels defined in terms of tolerable limits of impact or damage to the building or other structure, its contents and its occupants: mild, moderate, high and severe.

[BG] 304.2.1 Mild impact or damage. The tolerable impacts of the design loads are assumed as follows:

[BG] 304.2.1.1 Structural damage. The building or other structure does not have structural damage and is safe to occupy.

[BG] 304.2.1.2 Nonstructural systems. Nonstructural systems needed for normal building or other structure use and emergency operations are fully operational.

[BG] 304.2.1.3 Occupant hazards. Injuries to building or other structure occupants from hazard-related applied loads are minimal in numbers and minor in nature. There is a very low likelihood of single or multiple life loss. The nature of the applied load, such as fire hazards, may result in higher levels of expected injuries and damage in localized areas, whereas the balance of the areas may sustain fewer injuries and less damage.

[BG] 304.2.1.4 Overall extent of damage. Damage to building or other structure contents from hazard-related applied loads is minimal in extent and minor in cost.

[BG] 304.2.1.5 Hazardous materials. There is no significant release of hazardous materials outside of the building or other structure.The risk to the community is minimal, and an emergency relocation or shelter in place order is not necessary.

[BG] 304.2.2 Moderate impact or damage. The tolerable impacts of the design loads are assumed as follows:

[BG] 304.2.2.1 Structural damage. There is moderate structural damage, which is repairable; some delay in reoccupancy can be expected.

[BG] 304.2.2.2 Nonstructural systems. Nonstructural systems needed for normal building or other structure use are fully operational, although some cleanup and repair may be needed. Emergency systems remain fully operational.

[BG] 304.2.2.3 Occupant hazards. Injuries to building or other structure occupants from hazard-related applied loads may be locally significant, but generally moderate in numbers and in nature. There is a low likelihood of single life loss with a very low likelihood of multiple life loss. The nature of the applied load, such as fire hazards, may result in higher levels of expected injuries and damage in localized areas, whereas the balance of the areas may sustain fewer injuries and less damage.

[BG] 304.2.2.4 Overall extent of damage. Damage to building or other structure contents from hazard-related applied loads may be locally significant, but is generally moderate in extent and cost. The nature of the applied load, such as fire hazards, may result in higher levels of expected injuries and damage in localized areas, whereas the balance of the areas may sustain fewer injuries and less damage.

[BG] 304.2.2.5 Hazardous materials. There is no significant release of hazardous materials outside of the building or other structure. The risk to the community is minimal and an emergency relocation or shelter in place order is not necessary.

[BG] 304.2.3 High impact or damage. The tolerable impacts of the design loads are assumed as follows:

[BG] 304.2.3.1 Structural damage. There is significant damage to structural elements but there is not large falling debris; repair is possible. Significant delays in reoccupancy can be expected.

[BG] 304.2.3.2 Nonstructural systems. Nonstructural systems needed for normal building or other structure use are significantly damaged and inoperable; egress routes may be impaired by light debris; emergency systems may be significantly damaged, but remain operational.

[BG] 304.2.3.3 Occupant hazards. Injuries to building or other structure occupants from hazard-related applied loads may be locally significant with a high risk to life, but are generally moderate in numbers and in nature. There is a moderate likelihood of single life loss, with a low probability of multiple life loss. The nature of the applied load, such as fire hazards, may result in higher levels of expected injuries and damage in localized areas, whereas the balance of the areas may sustain fewer injuries and less damage.

[BG] 304.2.3.4 Overall extent of damage. Damage to building or other structure contents from hazard-related applied loads may be locally total and generally significant. The nature of the applied load, such as fire hazards, may result in higher levels of expected injuries and damage in localized areas, whereas the balance of the areas may sustain fewer injuries and less damage.

[BG] 304.2.3.5 Hazardous materials. There is no major release of hazardous materials outside of the facility. The risk to the community is minimal, and an emergency relocation or shelter in place order is not necessary.

[BG] 304.2.4 Severe impact or damage. The tolerable impacts of the design loads are assumed as follows:

[BG] 304.2.4.1 Structural damage. There is substantial structural damage, but all significant components continue to carry gravity load demands. Repair may not be technically possible. The building or other structure is not safe for application of loads, as reoccupancy could cause collapse.

[BG] 304.2.4.2 Nonstructural systems. Nonstructural systems for normal building or other structure use may be completely nonfunctional. Egress routes may be impaired; emergency systems may be substantially damaged and nonfunctional.

[BG] 304.2.4.3 Occupant hazards. Injuries to building or other structure occupants from hazard-related applied loads may be high in numbers and significant in nature. Significant risk to life may exist. There is a high likelihood of single life loss and a moderate likelihood of multiple life loss. The nature of the applied load, such as fire hazards, may result in higher levels of expected injuries and damage in localized areas, whereas the balance of the areas may sustain fewer injuries and less damage.

[BG] 304.2.4.4 Overall extent of damage. Damage to building or other structure contents from hazard-related applied loads may be total. The nature of the applied load, such as fire hazards, may result in higher levels of expected injuries and damage in localized areas, whereas the balance of the areas may sustain fewer injuries and less damage.

[BG] 304.2.4.5 Hazardous materials. Hazardous materials are released outside of the building or facility and an emergency relocation or shelter in place order may be needed.

SECTION 305—MAGNITUDES OF EVENT

Scan for Changes 5211670

[BG] 305.1 General. Magnitude of event encompasses all loads that can be reasonably expected to impact on a building or other structure, its users and its contents, during construction and throughout its intended life. This includes building and facility-related and occupancy-related loads, as well as loads resulting from natural and technological hazards.

Determination of magnitude of event shall take into account the design performance levels established by this code, the risk factors identified in Section 302.4 and specific performance criteria established by relevant authoritative documents.

[BG] 305.1.1 Natural hazards. The types of loads affecting main-force-resisting systems, components and contents that may be reasonably expected to impact on the building or other structure, its users and its contents during its intended life are provided in Chapter 5.

[BG] 305.1.2 Technological hazards. The types of loads due to technological hazards that may be reasonably expected to impact on the building or other structure, its users and its contents during construction and throughout its intended life include, but are not limited to:

[BG] 305.1.2.1 Fires (Chapter 15).

[BG] 305.1.2.2 Explosions (Chapters 5 and 19).

[BG] 305.1.2.3 Toxic materials (Chapter 19).

[BG] 305.1.2.4 Corrosive materials (Chapter 19).

[BG] 305.1.2.5 Infectious materials or agents (Chapter 19).

[BG] 305.2 Definition of magnitude of event. Magnitude of event can be defined, quantified and expressed either deterministically or probabilistically in accordance with the best current practice of the relevant profession as published in recognized authoritative documents. In some authoritative documents, magnitude of event may be expressed only for a single performance group. In other cases, magnitude of event may be provided for all performance levels such as seismic provisions. In all cases, it is the responsibility of the design engineer to demonstrate that the design performance levels are met for the loads anticipated.

[BG] 305.2.1 Classification of event magnitude. For the purpose of this code, the magnitude of event shall be classified as: small, medium, large and very large. Where authoritative documents do not present magnitude of event in this format, it will be the responsibility of the designer to relate the loads to this format and to demonstrate that the minimum design performance levels will be met by the proposed design.

CHAPTER 4

RELIABILITY AND DURABILITY

User notes:

About this chapter: *Chapter 4 underscores the importance of the reliability of individual protection systems and strategies, as well as the reliability of the interaction of these systems in achieving the design performance level for a particular building or facility addressed in Chapter 3. Reliability is a function of many factors, including redundancy, maintenance, durability of materials, quality of installations and integrity of design.*

SECTION 401—RELIABILITY

[BG] 401.1 Objective. To ensure reliability of the system necessary to meeting the performance objectives of building, facility or processes in accordance with the design.

[BG] 401.2 Functional statements.

[BG] 401.2.1 Design, installation and maintenance. Design, install and maintain systems, system components and equipment that provide a safety function in strict accordance with the manufacturers' instructions and with any applicable codes and standards.

[BG] 401.2.2 Testing and inspection. Test and inspect systems, system components and equipment that provide a safety function in strict accordance with the manufacturers' instructions and with any applicable codes and standards for both the methods employed and the frequency.

[BG] 401.2.3 Active fire protection systems. Active fire protection systems such as fire alarm, suppression and smoke management systems shall undergo *commissioning* testing when first placed into service or following any substantial alteration.

[BG] 401.2.4 Training. Provide appropriate training to any people who operate, test, maintain or interpret information from any safety systems. Where such work is done by contractors, ensure that they have the necessary training and skills.

[BG] 401.3 Performance requirements.

[BG] 401.3.1 Qualifications. Design, installation and maintenance shall be performed only by qualified people as approved. Certification or records of training shall be provided.

[BG] 401.3.2 Documentation. Documentation shall be maintained at the building that details the systems installed and their required maintenance and testing methods and frequency. Records of such maintenance and testing shall be maintained that demonstrate compliance, the persons conducting the work and their qualifications.

SECTION 402—DURABILITY

[BG] 402.1 Objective. To assist in the selection of appropriate materials and construction systems.

[BG] 402.2 Functional statement. To ensure that a building will continue to satisfy the objectives of this code throughout its life.

[BG] 402.3 Performance requirements.

[BG] 402.3.1 Normal maintenance. From the time a certificate of occupancy is issued, primary building elements shall, with only normal maintenance, continue to satisfy the performance requirements of this code for the intended life of the building.

[BG] 402.3.2 Intended life of a building. Where the useful life of building or facility elements or systems is less than the intended life of the building, provisions shall be made for timely replacement of those elements, so that the objectives of this code and the design are maintained.

[BG] 402.3.3 Damage and deterioration. Where damage or deterioration to building or facility elements or systems will impact the objectives of this code or the design, those elements or systems shall be repaired or replaced in order to maintain the level of performance intended by this code.

[BG] 402.3.4 Determination of durability and service life. In determining the useful service life of building elements, products or systems, an acceptable method for determining durability and service life shall be used.

CHAPTER 5 STABILITY

User notes:

About this chapter: *Chapter 5 provides the requirements for the structural design of buildings and other structures. Section 501 specifies the forces for which structures need to be designed and the required performance. This chapter requires a structure to be designed for the expected forces it will be subjected to throughout its life. This is the same requirement found in Chapter 16 of the* International Building Code.

Code development reminder: *Code change proposals to this appendix will be considered by the IBC—General Code Development Committee during the 2025 (Group B) Code Development Cycle.*

SECTION 501—STRUCTURAL FORCES

Scan for Changes c4715e7

[BS] 501.1 Objective. To provide a desired level of structural performance when structures are subjected to the loads that are expected during construction or alteration and throughout their intended lives.

[BS] 501.2 Functional statements.

[BS] 501.2.1 Life safety and injury prevention. Structures shall be designed and constructed to prevent injury to occupants due to loading of a structural element or system consistent with the design performance level determined in Chapter 3.

[BS] 501.2.2 Property and amenity protection. Structures shall be designed and constructed to prevent loss of property and *amenity* consistent with the design performance level determined in Chapter 3.

[BS] 501.3 Performance requirements. Performance of the structure shall be determined based on loads derived to satisfy the requirements of this section. Where the provisions of this section are not used to develop loads, loads shall be determined in accordance with Chapter 16 of the *International Building Code*.

[BS] 501.3.1 Stability. Structures, or portions thereof, shall remain stable and not collapse during construction or alteration and throughout their lives.

[BS] 501.3.2 Disproportionate failure. Structures shall be designed to sustain local damage, and the structural system as a whole shall remain stable and not be damaged to an extent disproportionate to the original local damage.

[BS] 501.3.3 Loss of amenity. Structures, or portions thereof, shall have a low probability of causing damage or loss of *amenity* through excessive deformation, vibration or degradation during construction or alteration and throughout their lives.

[BS] 501.3.4 Expected loads. Structures, or portions thereof, shall be designed and constructed taking into account expected loads, and combination of loads, associated with the event(s) magnitude(s) that would affect their performance, including, but not limited to:

1. Dead loads.
2. Live loads.
3. Impact loads.
4. Explosion loads.
5. Soil and hydrostatic pressure loads.
6. Flood loads (mean return period).
   - 6.1. Small: 100 years.
   - 6.2. Medium: 500 years.
   - 6.3. Large: Determined on a site-specific basis.
   - 6.4. Very Large: Determined on a site-specific basis.
7. Wind loads (mean return period).
   - 7.1. Small: 300 years.
   - 7.2. Medium: 700 years.
   - 7.3. Large: 1,700 years.
   - 7.4. Very Large: 3,000 years.
8. Windborne debris loads.
9. Snow loads (mean return period).
   - 9.1. Small: 25 years.
   - 9.2. Medium: 50 years.
   - 9.3. Large: 100 years.

9.4. Very Large: 500 years.
9.5. Snow loads shall Include but not be limited to consideration for drifting, unbalanced loads, sliding snow loads and ice damming.

10. Rain loads. See Table 501.3.4.
11. Earthquake loads.
    11.1. Small: 43 years (mean return period).
    11.2. Medium: 72 years (mean return period).
    11.3. Large: Two-thirds of the intensity of very large loads.
    11.4. Very large: The Risk-Targeted Maximum Considered Earthquake defined in Chapter 21 of ASCE 7.
12. Ice loads, atmospheric icing (mean return period).
    12.1. Small: 25 years.
    12.2. Medium: 50 years.
    12.3. Large: 100 years.
    12.4. Very Large: 200 years.
13. Hail loads.
14. Thermal loads.
15. Loads due to Tsunamis.
    15.1. Very Large: The Maximum Considered Tsunami Defined in Chapter 6 of ASCE 7.

[BS] TABLE 501.3.4—RAIN LOADS

MAGNITUDE OF EVENT	DRAINAGE SYSTEM	MRI (years)	STORM DURATION (min.)
Small	Primary	25	60
Small	Secondary	25	15
Medium	Primary	50	60
Medium	Secondary	50	15
Large	Primary	100	60
Large	Secondary	100	15
Very Large	Primary	100	30
Very Large	Secondary	100	10

[BS] 501.3.5 Target reliabilities. The design of buildings and structures shall consider appropriate factors of safety to provide the target reliabilities stipulated in Chapter 1 of ASCE 7 considering:

1. Effects of uncertainties resulting from construction activities.
2. Variation in the properties of materials and the characteristics of the site.
3. Accuracy and limitations inherent in the methods used to predict the stability of the building, load effects and capacities of members and their connections.
4. Self-straining forces arising from differential settlements of foundations and from restrained dimensional changes due to temperature, moisture, shrinkage, creep and similar effects.
5. Uncertainties in the determination of the expected loads.
6. The level of quality control and *quality assurance* in construction.

[BS] 501.3.6 Demolition and alteration. The demolition or alteration of buildings and structures shall be carried out in a way that avoids the likelihood of premature collapse.

[BS] 501.3.7 Site work. Site work, where necessary, shall be carried out to provide stability for construction on the site and avoid the likelihood of damage to adjacent property.

CHAPTER

6 PEDESTRIAN CIRCULATION

User notes:

About this chapter: *Chapter 6 provides performance requirements for three aspects of pedestrian circulation in a building:*

1. Means of egress: Section 601 provides guidance by which egress systems for buildings and facilities are designed, evaluated and maintained.
2. Accessibility: Section 602 provides guidelines to ensure disabled persons have reasonable access to and use of buildings to an extent consistent with that to which people without disabilities are able to access and use buildings. This is consistent with the intent of federal statutes enacted to protect the rights of people with disabilities.
3. Transportation equipment: Section 603 provides general safety guidelines for installation of elevators, dumbwaiters and escalators inside or outside buildings. These provisions include use during normal operations, use by firefighters during emergency operations and use by maintenance personnel during activities associated with adjusting, servicing and inspecting elevators.

SECTION 601—MEANS OF EGRESS

[BE] 601.1 Objective. To protect people during egress and rescue operations.

[BE] 601.2 Functional statement. Enable occupants to exit the building, facility and premises or reach a *safe place* as appropriate to the design performance level determined in Chapter 3.

[BE] 601.3 Performance requirements.

[BE] 601.3.1 General. The construction, arrangement and number of means of egress, exits and safe places for buildings shall be appropriate to the travel distance, number of occupants, occupant characteristics, building height, and *safety systems* and features.

[BE] 601.3.2 Identification, illumination and safety of means of egress. Means of egress shall be clearly identified, provided with adequate illumination and be easy and safe to use.

[BE] 601.3.3 Unobstructed path. Means of egress shall provide an unobstructed path of travel from each *safe place* to not less than one exit.

[BE] 601.3.4 Protection from untenable conditions. Each *safe place* shall provide adequate protection from untenable conditions, an appropriate communication system and adequate space for the intended occupants.

[BE] 601.3.5 Human biomechanics and expectation of consistency. Means of egress shall enable reasonable use by the occupants in the building with due regard to human biomechanics and expectation of consistency.

[BE] 601.3.6 Maintenance of means-of-egress systems. Suitable means of egress shall be provided in satisfactory arrangement throughout all buildings, facilities and premises, regardless of when they were constructed, based on the number and character of occupants, length of travel, provision of existing alternative paths, timeline of emergency detection and response, risk level, time to exit and *safety systems* provided.

[BE] 601.3.7 Maintenance of clear path. Means of egress shall be maintained without obstructions or reductions in capacity that would hinder the ability of the occupants to egress safely.

[BE] 601.3.8 Interference with identification of exits. Means of egress shall be readily identifiable. Buildings shall be operated and maintained in a manner that does not interfere with the identification of exits.

[BE] 601.3.9 Ease of use. Means of egress shall be maintained and operated in such a manner to ensure that all egress facilities are readily openable and available without special knowledge or effort consistent with the use or occupancy characteristics.

[BE] 601.3.10 Maintenance of illumination. Means of egress shall be maintained and operated in such a manner to ensure that adequate lighting to facilitate safe egress is available.

SECTION 602—ACCESSIBILITY

[BE] 602.1 Objective. To provide people with disabilities reasonable use of the built environment in a manner consistent with that provided to people without disabilities in nonemergency conditions.

[BE] 602.2 Functional statement. Buildings and their site-adjacent facilities shall allow all people, including but not limited to people with disabilities, functional use of spaces based on a space's intended purpose.

[BE] 602.3 Performance requirements. Safe and usable routes shall be provided that allow people to:

1. Approach, enter and leave buildings and sites to and from adjacent transportation stops, walkways and parking areas.
2. Move within and use building and site spaces based on a facility's intended purpose.

SECTION 603—TRANSPORTATION EQUIPMENT

[BG] 603.1 Objective. To ensure the safety of all people using, maintaining and inspecting elevators, escalators and similar building transportation equipment inside or outside of buildings.

[BG] 603.2 Functional statement. Building transportation equipment installations for access into, within and outside of buildings shall provide for the safe movement of all people and the safety of maintenance and inspection personnel.

[BG] 603.3 Performance requirements.

[BG] 603.3.1 General. Building transportation equipment shall:

1. Move people safely when starting, stopping, accelerating, decelerating or changing direction of travel, and hold the rated loads.
2. Be constructed to avoid the likelihood of people falling, tripping, becoming caught and coming in contact with sharp edges or projections under normal and reasonably foreseeable conditions of use.
3. Be guided and have sufficient running clearances.
4. Have controls to stop and prevent restarting in the event of activation of a safety device.
5. Be capable of being isolated for inspection, testing and maintenance.
6. Have adequate lighting and ventilation during normal conditions or upon loss of normal power.

[BG] 603.3.2 Elevators. Elevators shall be designed and constructed to provide:

1. A means of communication for trapped passengers in stalled elevators.
2. Emergency recall operation that discharges passengers at the required designated or alternate landing in the event of a fire emergency.
3. Emergency in-car operation for firefighting and rescue operations.
4. An environment that ensures the safe operation of the equipment for the anticipated use or application.

CHAPTER

7 SAFETY OF USERS

User notes:

About this chapter: *Chapter 7 addresses protection of people from hazards related to building materials, such as glazing or materials that emit radiation, and construction and demolition hazards. The chapter specifically provides performance requirements for signs and emergency notification. Hazardous materials are regulated in Chapter 19.*

SECTION 701—HAZARDS FROM BUILDING MATERIALS

[BG] 701.1 Objective. To safeguard people from injury caused by exposure to hazards from building materials.

[BG] 701.2 Functional statement. Building materials that are potentially hazardous shall be used in ways to avoid undue risk to people.

[BG] 701.3 Performance requirements.

[BG] 701.3.1 Construction materials. The quantities of gas, liquid, radiation or solid particles emitted by materials used in the construction of buildings shall not give rise to harmful concentrations at the surface of the material where the material is exposed or in the atmosphere of any space.

[BG] 701.3.2 Glazing. Glass or other brittle materials with which people are likely to come into contact shall comply with one or more of the following:

1. If broken upon impact, break in a way that is unlikely to cause injury.
2. Resist a reasonably foreseeable impact without breaking.
3. Be reasonably protected from impact.

SECTION 702—PREVENTION OF FALLS

[BG] 702.1 Objective. To prevent people from unintentionally falling from one level to another.

[BG] 702.2 Functional statement. Buildings and their facilities shall be constructed to reduce the likelihood of unintentional falls.

[BG] 702.3 Performance requirements.

[BG] 702.3.1 Required barriers. A barrier shall be provided where people could fall 30 inches (762 mm) or more from an opening in the external envelope or floor of a building or its facilities.

[BG] 702.3.2 Roofs. Roofs with permanent access shall have barriers provided.

[BG] 702.3.3 Barrier construction. Barriers shall be constructed and installed appropriate to the hazard.

[BG] 702.3.4 Openings in barriers. Where barriers have openings, the openings shall be of an appropriate size and configuration to keep people from falling through based on the anticipated age of the occupants.

SECTION 703—CONSTRUCTION AND DEMOLITION HAZARDS

[BG] 703.1 Objective. To safeguard people from injury or illness and to protect property from damage during the construction or demolition processes.

[BG] 703.2 Functional statement. Provisions are required during construction and demolition work to:

1. Protect authorized personnel from injury resulting from falling objects, fire, blasts, tripping or falling, or any other risk posed by the construction or demolition operation.
2. Prevent the entry of unauthorized personnel on the construction or demolition site.
3. Protect property off site from damage resulting from falling objects, fire, blasts or any other risk posed by the construction or demolition operations.

[BG] 703.3 Performance requirements.

[BG] 703.3.1 Operations and procedures. Sequencing of tasks, procedural methods and equipment shall be such that:

1. Personnel are protected from injury and illness attributable to hazards present because of the given operation.
2. Adjacent property, property on site and equipment are protected from damage from execution of the given tasks.
3. Safety procedures limit the accumulation of combustible materials on site and provide safeguards for equipment and operations that represent ignition sources.

[BG] 703.3.2 Protection from natural hazards. The structure under construction shall be protected from damage due to wind, rain or other natural hazards likely to occur during construction.

[BG] 703.3.3 Protection of personnel. Provisions for personnel movement, transport and support shall be such that:

1. Personnel are protected from injury due to falling.
2. Personnel are protected from injury due to falling objects.

3. Personnel are protected from injury that could be caused by the particular operations being conducted.
4. Exposure to materials that are known to be health hazards is eliminated.

[BG] 703.3.4 Unauthorized entry. The job site shall be protected from the intrusion of unauthorized personnel.

SECTION 704—SIGNS

[F] 704.1 Objective. To identify essential features of the building to its users.

[F] 704.2 Functional statement. Signs shall identify escape and rescue routes, hazards, accessible elements where not all elements are accessible and other essential features of a building.

[F] 704.3 Performance requirements.

[F] 704.3.1 Visibility. Signs shall be clearly visible and readily recognizable under the conditions expected for their purpose.

[F] 704.3.2 Identification of exits and safe places. Signs shall identify exits and safe places, and be located sufficiently to mark escape/rescue routes and guide people to exits and safe places.

[F] 704.3.3 Power failure. Signs that identify exits, safe places and escape/rescue routes shall remain visible in the event of a power failure.

[F] 704.3.4 Hazard identification. Signs indicating hazards shall be provided in sufficient locations to notify people before they encounter the hazard.

[F] 704.3.5 Accessible building feature signage. Signs shall identify accessible facilities and be located sufficiently to mark accessible routes.

SECTION 705—EMERGENCY NOTIFICATION

[F] 705.1 Objective. To provide notification of the need to take some manual action to preserve the safety of occupants or to limit property damage.

Scan for Changes 3e4ba8e

[F] 705.2 Functional statements.

[F] 705.2.1 Occupant notification. Where required in accordance with the design report per Section 102.3.4.2.2, adequate means of occupant notification shall be provided to warn of the presence of a fire or other emergency in sufficient time to enable occupants to take the contemplated action without being exposed to unreasonable risk of injury or death.

[F] 705.2.2 Emergency responder notification. Where systems are designed to notify emergency responders, such systems shall indicate the type of emergency and the location of the buildings, premises or other structures. Where such buildings, premises or other structures are large enough to expect difficulty in prompt location of the fire or other public emergency, identification of the area or zone of the emergency shall be provided at the buildings, premises or other structures.

[F] 705.3 Performance requirements.

[F] 705.3.1 Type of notification. Notification of occupants shall be by means appropriate to the needs of the occupants, the use of the building and the emergency egress strategy employed.

[F] 705.3.2 Sleeping occupants. Where required by the anticipated use of the building, notification systems shall be capable of alerting sleeping occupants in reasonable time to enable them to reach a *safe place* before the occurrence of untenable conditions at any point along the primary egress path.

CHAPTER

8 MOISTURE

User notes:

About this chapter: *Chapter 8 is intended to prevent moisture from adversely affecting occupant health and safety and the structural and functional performance of a building. This chapter deals with both the liquid and vapor forms of water.*

SECTION 801—SURFACE WATER

[BS] 801.1 Objective. To safeguard people from injury and protect the building or other property from damage caused by surface water and to protect outfalls of drainage systems that may become contaminated from on-site hazardous material storage.

[BS] 801.2 Functional statements.

[BS] 801.2.1 Surface water hazards. Buildings and sites shall be constructed in a way that protects people and other property from the adverse effects of surface water.

[BS] 801.2.2 Hazardous materials contamination. Building or building sites used for the storage or use of hazardous materials shall include provisions to ensure that hazardous materials are not accidentally transported across property lines into drainage outfalls.

[BS] 801.3 Performance requirements.

[BS] 801.3.1 Removal of surface water. Surface water shall be removed in a manner that avoids damage or nuisance to the building or other property.

[BS] 801.3.2 Contaminated water. Surface water or water used for firefighting and other uses shall be routed so as not to transport hazardous material across property lines or into drainage outfalls.

[BS] 801.3.3 Surface water runoff. Surface water drainage systems shall convey surface water runoff to an appropriate outfall.

[BS] 801.3.4 Blockage. Drainage systems shall be constructed so as to avoid the likelihood of blockage.

[BS] 801.3.5 Access for cleaning. Drainage systems shall be constructed so as to have reasonable *access* for cleaning.

SECTION 802—EXTERNAL MOISTURE

Scan for Changes
95caf6a

[BS] 802.1 Objective. To safeguard people from injury and to protect the building, facilities, equipment, processes, materials, and contents from damage that could result from external moisture entering the building.

[BS] 802.2 Functional statement. Buildings shall be constructed to provide adequate resistance to penetration by, and the accumulation of, moisture from the outside.

[BS] 802.3 Performance requirements.

[BS] 802.3.1 Water penetration. Roofs and exterior walls shall prevent the unwanted penetration and accumulation of moisture or water that causes damage to the building, facilities, equipment, processes, materials or contents, and shall provide a means for unwanted penetration of water or moisture to dissipate.

[BS] 802.3.2 Building elements in contact with the ground. Walls, floors and structural support elements in contact with the ground shall not absorb or transmit moisture in quantities that could cause damage to the building facilities, equipment, processes, materials or contents.

[BS] 802.3.3 Concealed spaces and cavities. Concealed spaces and cavities in buildings shall be constructed in a way that prevents unwanted penetration and accumulation of moisture or water that causes damage to the building, facilities, equipment, processes, materials or contents, and shall provide a means for any unwanted penetration of water or moisture to dissipate without causing damage.

[BS] 802.3.4 Moisture during construction. Excess moisture present at the completion of construction shall be capable of being dissipated without permanent damage to building elements.

SECTION 803—INTERNAL MOISTURE

[BS] 803.1 Objective. To safeguard people against illness or injury that could result from accumulation of internal moisture, and to protect an occupancy from damage caused by free water from another occupancy in the same building.

[BS] 803.2 Functional statement. Buildings shall be constructed to avoid the likelihood of:

1. Fungal growths or the accumulation of contaminants on linings and other building elements.
2. Free water overflow penetrating to an adjoining occupancy.
3. Damage to building elements being caused by the use of water.

[BS] 803.3 Performance requirements.

[BS] 803.3.1 Excess moisture removal and protection. An adequate means shall be provided to remove excess moisture from the structure or protect it from the effects of excess moisture and condensation in all habitable spaces, bathrooms, laundries and other locations where moisture is potentially generated.

[BS] 803.3.2 Overflow. Accidental overflow from sanitary fixtures or laundering facilities shall be constrained from penetrating another occupancy in the same building.

[BS] 803.3.3 Floor surfaces. Floor surfaces of any space containing sanitary fixtures or laundering facilities shall be impervious to water and easily cleaned.

[BS] 803.3.4 Wall surfaces. Wall surfaces adjacent to sanitary fixtures or laundering facilities shall be impervious to water and easily cleaned.

[BS] 803.3.5 Surfaces and building elements. Surfaces of building elements likely to be splashed or to become contaminated in the course of the intended use of the building shall be impervious to water and easily cleaned.

[BS] 803.3.6 Water splash. Water splash shall be prevented from penetrating behind linings or into concealed spaces.

CHAPTER

9 INTERIOR ENVIRONMENT

User notes:

About this chapter: *Chapter 9 describes the performance provisions of a building's environment. It is divided into four areas:*

1. *Section 901 addresses the space and climate appropriate for the needs of the occupants, the activities and the furnishings.*
2. *Section 902 addresses the need to provide adequate clean air in a building or facility. This includes controlling the moisture content, odors, poisonous fumes, etc. This section is aimed at such problems as "sick building syndrome."*
3. *Section 903 requires that tenant, common and habitable spaces be insulated against sound transmission.*
4. *Section 904 establishes the criteria for light for everyday use in habitable spaces and means of egress.*

SECTION 901—CLIMATE AND BUILDING FUNCTIONALITY

[BG] 901.1 Objective. To safeguard people from illness caused by air temperature and to safeguard people from injury or loss of *amenity* caused by inadequate activity space.

[BG] 901.2 Functional statements. Buildings shall be constructed to provide:

1. Adequately controlled interior temperatures.
2. Adequate activity space for the intended use.

[BG] 901.3 Performance requirements.

[BG] 901.3.1 Temperature. Habitable spaces, bathrooms and recreation rooms shall be designed to maintain the internal temperature at a level sufficient for the occupants while the space is adequately ventilated.

[BG] 901.3.2 Space. Habitable spaces shall have sufficient space for activity, furniture and sanitary needs of the occupants.

SECTION 902—INDOOR AIR QUALITY

[BG] 902.1 Objective. To maintain the habitable spaces of buildings and facilities with an environment that is conducive to the comfort, health and safety of the occupants.

[BG] 902.2 Functional statement. Habitable spaces within buildings shall be provided with air that contains sufficient oxygen and limits the levels of moisture and contaminants to levels that are consistent with good health, safety and comfort.

[BG] 902.3 Performance requirements.

[BG] 902.3.1 Ventilation. Habitable spaces within buildings shall have a means of ventilation that maintains air quality at all times that the spaces are occupied and with the maximum number of occupants anticipated.

[BG] 902.3.2 Collection and removal. Buildings shall have a means of collecting or otherwise removing the following products from the spaces in which they are generated:

1. Cooking fumes and odors.
2. Excessive water vapor from laundering, utensil washing, bathing and showering.
3. Odors from sanitary and waste storage spaces.
4. Gaseous byproducts and excessive moisture from commercial or industrial processes.
5. Poisonous fumes and gases.
6. Airborne particles.
7. Products of combustion.
8. Off-gases from building materials, fixtures and contents.

[BG] 902.3.3 Building materials. Building materials that release quantities of contaminants that cannot be maintained at safe levels shall not be used.

[BG] 902.3.4 Contaminated air. Contaminated air shall be disposed of in a way that avoids creating a nuisance or hazard to people and other property.

[BG] 902.3.5 Sufficient supply air. The quantity of air supplied for ventilation shall account for the demands of any fixed combustion appliances.

SECTION 903—AIRBORNE AND IMPACT SOUND

[BG] 903.1 Objective. To safeguard people from loss of *amenity* as a result of excessive noise being transmitted between adjacent tenants or occupancies.

[BG] 903.2 Functional statement. Building elements that are common between tenants or occupancies shall be constructed to prevent excessive noise transmission from other tenants or occupancies or common spaces to habitable spaces.

[BG] 903.3 Performance requirements.

[BG] 903.3.1 Tenant separations. The airborne transmission of sound through tenant separation walls and floors shall be reduced to a level that minimizes its effect on adjacent occupants.

[BG] 903.3.2 Floors. The structure-borne transmission of sound through floors shall be reduced to a level that minimizes its effect on adjacent occupants.

SECTION 904—ARTIFICIAL AND NATURAL LIGHT

[BG] 904.1 Objective. To safeguard people from injury or loss of *amenity* due to lack of adequate lighting.

[BG] 904.2 Functional statements.

[BG] 904.2.1 Lighting for safe movement. Habitable spaces and means of egress within buildings shall be provided with adequate artificial lighting to enable safe movement.

[BG] 904.2.2 General lighting. Adequate natural or artificial light shall be provided in all habitable spaces.

[BG] 904.3 Performance requirements.

[BG] 904.3.1 Illumination. Adequate illumination shall be provided appropriate to the use and occupancy of the habitable spaces and means of egress served.

[BG] 904.3.2 Natural light. Natural light shall provide a luminance appropriate to the use and occupancy of the habitable spaces served.

CHAPTER

10 MECHANICAL

User notes:

About this chapter: *Chapter 10 describes the performance provisions for building mechanical systems. It addresses the installation and functions of HVAC, refrigeration and piped services. The provisions of this chapter focus on the performance of the equipment versus issues related to indoor air quality. Chapter 9 provides more guidance on indoor air quality and climate.*

SECTION 1001—HEATING, VENTILATION AND AIR-CONDITIONING EQUIPMENT (HVAC)

[M] 1001.1 Objective. To provide the safe installation of the equipment to condition the air for the health and comfort of the occupants.

[M] 1001.2 Functional statement. The installation of equipment shall safeguard maintenance personnel and building occupants from injury and deliver air at the appropriate temperature for health and comfort.

[M] 1001.3 Performance requirements.

[M] 1001.3.1 Protection from equipment. People and building elements shall be protected from contact with hot and live electrical parts.

[M] 1001.3.2 Service and replacement ability. The HVAC system shall allow safe isolation and *access* for service and replacement of equipment.

[M] 1001.3.3 Temperature controls. The HVAC system shall include devices to monitor and control the temperature.

[M] 1001.3.4 Securing of equipment. HVAC equipment and appliances shall be secured in place.

SECTION 1002—REFRIGERATION

[M] 1002.1 Objective. To provide the safe installation and operation of refrigeration equipment.

[M] 1002.2 Functional statement. The installation of equipment shall safeguard maintenance personnel and building occupants from injury.

[M] 1002.3 Performance requirements.

[M] 1002.3.1 Protection from equipment. People and building elements shall be protected from contact with hot or live electrical parts.

[M] 1002.3.2 Service and replacement ability. Refrigeration equipment shall allow safe isolation and *access* for service and replacement of equipment.

[M] 1002.3.3 Temperature controls. Refrigeration equipment shall include devices to monitor and control temperature.

[M] 1002.3.4 Toxic and flammable refrigerants. Refrigeration equipment shall have appropriate safeguards where utilizing toxic or flammable refrigeration agents.

SECTION 1003—PIPED SERVICES

[M] 1003.1 Objective. To safeguard people from injury or illness caused by extreme temperatures or hazardous substances associated with building services.

[M] 1003.2 Functional statement. In buildings with potentially hazardous services containing hot, cold, flammable, corrosive or toxic liquids or gases, the installations shall be constructed to provide adequate safety for people.

[M] 1003.3 Performance requirements.

[M] 1003.3.1 Construction. Piping systems shall be constructed to avoid the likelihood of:

1. Significant leakage or damage during normal or reasonably foreseeable abnormal conditions.
2. Detrimental contamination of the contents by other substances.
3. Adverse interaction between services or between piping and electrical systems.
4. People having contact with pipes that could cause them harm.

[M] 1003.3.2 Corrosion. Pipes shall be protected against corrosion in the environment of their use.

[M] 1003.3.3 Identification. Piping systems shall be identified with markings if the contents are not readily apparent from the location or associated equipment.

[M] 1003.3.4 Enclosed spaces. Enclosed spaces shall be constructed to avoid the likelihood of accumulating vented or leaking flammable gas.

[M] 1003.3.5 Isolation. A piped system shall have isolation devices that permit the complete system or components of the system to be isolated from the supply system for maintenance, testing, fault detection and repair.

CHAPTER

11 PLUMBING

User notes:

About this chapter: *Chapter 11contains performance provisions regarding facilities for personal hygiene, laundering, domestic water supplies and wastewater. Accessibility must be addressed for these provisions based on the application of Section 602. The provisions for personal hygiene require that proper sanitary fixtures, and such things as adequate toilet facilities, be provided for the occupants of a building to maintain health and to prevent the spread of disease.*

SECTION 1101—PERSONAL HYGIENE

[P] 1101.1 Objective. To provide facilities with appropriate space, fixtures and equipment for personal hygiene.

[P] 1101.2 Functional statement. To provide adequate plumbing fixtures that reasonably protect people from illness and provide reasonable access to such fixtures conducive to health, safety and comfort of the occupants.

[P] 1101.3 Performance requirements.

[P] 1101.3.1 Number of plumbing fixtures. Plumbing fixtures shall be provided in sufficient numbers appropriate for the intended use.

[P] 1101.3.2 Privacy. Plumbing fixtures shall be located to provide appropriate privacy.

[P] 1101.3.3 Cleanliness. Plumbing fixtures shall be constructed to avoid food contamination and accumulation of dirt or bacteria, and permit effective cleaning.

[P] 1101.3.4 Wastewater removal. Plumbing fixtures shall be installed to discharge to drainage systems without contaminating food.

[P] 1101.3.5 Location of plumbing fixtures. Facilities for personal hygiene shall be provided in convenient locations and spaces of appropriate size to permit the use of the fixtures.

SECTION 1102—LAUNDERING

[P] 1102.1 Objective. To provide adequate facilities for laundry.

[P] 1102.2 Functional statement. Laundry facilities shall be provided for use by occupants of dwelling units.

[P] 1102.3 Performance requirement. Space shall be adequate in size for the required fixtures and equipment.

SECTION 1103—DOMESTIC WATER SUPPLIES

[P] 1103.1 Objective. To provide sanitary distribution of water for drinking, food preparation and hygiene.

[P] 1103.2 Functional statement. Sanitary water shall be delivered to fixtures, appliances and equipment at temperatures appropriate for the intended use.

[P] 1103.3 Performance requirements.

[P] 1103.3.1 Potable water. Water supplies intended for human consumption, oral hygiene, food preparation and the washing of cooking equipment shall be potable.

[P] 1103.3.2 Hot water. Plumbing fixtures and appliances used for personal hygiene, laundering and the washing of cooking equipment shall be provided with hot water.

[P] 1103.3.3 Scalding. Where hot water is provided for personal hygiene, it shall be delivered at a temperature to avoid scalding.

[P] 1103.3.4 Water supply contamination. Water supplies shall be installed to avoid potable water contamination.

[P] 1103.3.5 Flow rate and pressure. Water supplies shall be provided to plumbing fixtures, appliances and equipment at a flow rate and pressure adequate for their operation.

[P] 1103.3.6 Leak prevention. Water piping shall be installed in a leak-free manner.

[P] 1103.3.7 Access. Water systems shall be installed to allow adequate *access* for maintenance.

[P] 1103.3.8 Water piping isolation and protection from contamination. Water piping shall be installed with provisions for adequate isolation of the system and branches and to provide protection from contamination.

[P] 1103.3.9 Hot water vessels. Vessels used for producing hot water shall be provided with safety devices to relieve excessive pressure and limit temperatures.

SECTION 1104—WASTEWATER

Scan for Changes 3a8baad

[P] 1104.1 Objective. To provide safe drainage and disposal systems for wastewater from plumbing fixtures, appliances and equipment.

[P] 1104.2 Functional statement. The drainage system shall conduct wastewater to an appropriate disposal point, protect people from contamination and unpleasant odor, and avoid blockages.

[P] 1104.3 Performance requirements.

[P] 1104.3.1 Prevention of blockage and leakage. The drainage system shall conduct wastewater from all plumbing fixtures, appliances and equipment, avoiding the likelihood of blockage and leakage.

[P] 1104.3.2 Sewer gases. The drainage system shall be designed and installed to prevent sewer gases from entering the building.

[P] 1104.3.3 Access. The drainage system shall have *access* for maintenance and clearing of blockages.

[P] 1104.3.4 Sewer connection. The drainage system shall be connected to the sewer in a manner acceptable to the operator of the sewer system.

[P] 1104.3.5 On-site sewage disposal. On-site sewage disposal systems shall be designed and installed in an approved manner.

SECTION 1105—NONPOTABLE WATER SYSTEMS

Scan for Changes d873404

[P] 1105.1 Objective. To provide safe nonpotable water systems consisting of collection, storage, treatment and distribution components for graywater, rainwater and recycled water sources.

[P] 1105.2 Functional statement. Nonpotable water systems shall collect nonpotable water from approved sources, provide treatment and storage of nonpotable water, and distribute nonpotable water to approved fixtures and outlets.

[P] 1105.3 Performance requirements. The performance requirements of nonpotable water systems shall be in accordance with Sections 1105.3.1 through 1105.3.7.2.

[P] 1105.3.1 Identification. Nonpotable water systems shall be clearly identified.

[P] 1105.3.2 Separation. Nonpotable water systems shall be provided in systems isolated from potable water systems to avoid potable water contamination.

[P] 1105.3.3 Water quality. Nonpotable water shall meet the minimum water quality requirements established by the jurisdiction.

[P] 1105.3.4 Flow rate and pressure. Nonpotable water shall be supplied to fixtures and outlets at a flow rate and pressure adequate for their operation.

[P] 1105.3.5 Leak prevention. Piping and storage tanks for nonpotable water systems shall be installed in a leak-free manner.

[P] 1105.3.6 Access. Nonpotable water systems shall be installed to allow adequate *access* for maintenance.

[P] 1105.3.7 Storage. Storage of nonpotable water collected on-site shall be provided of sufficient size and capacity to support the intended uses.

[P] 1105.3.7.1 Venting and overflow. Nonpotable water storage shall be designed and installed to allow venting of gases and to control overflow without damage to the system, in accordance with requirements established by the jurisdiction.

[P] 1105.3.7.2 Makeup water. Where required for the intended uses, other sources of water supply shall be provided to ensure that there is an uninterrupted supply of water from the nonpotable water system.

CHAPTER

12 FUEL GAS

User notes:

About this chapter: *Chapter 12 provides requirements to ensure that use of fuel gas does not create a hazard to occupants. This primarily means that the fuel gas shall not produce unsafe levels of combustible byproducts or become a source of ignition of an unwanted fire.*

SECTION 1201—FUEL GAS PIPING AND VENTS

[FG] 1201.1 Objective. To ensure that fuel gas is distributed and utilized in a safe manner.

[FG] 1201.2 Functional statement. In buildings where fuel gas is used as an energy source, the vented and unvented gas piping systems shall be safe and adequate for their intended use.

[FG] 1201.3 Performance requirements.

[FG] 1201.3.1 General. Gas piping systems shall be free of leaks and operated at a safe pressure appropriate to the appliances served by the system.

[FG] 1201.3.2 Isolation. Gas piping systems shall have isolation devices that permit isolation of appliances, or isolation of the gas piping systems from the supply, for maintenance, testing, leak detection or repair.

[FG] 1201.3.3 Conveyance of products of combustion. Vented gas appliances shall convey products of combustion directly to the exterior without affecting the operation of other gas vents.

[FG] 1201.3.4 Safety controls. Vented gas appliances shall be provided with safety controls that prevent their operation in the event of failure of forced ventilation systems or natural draft systems.

CHAPTER

13 ELECTRICITY

User notes:

About this chapter: *Chapter 13 provides for the installation of electrical services and equipment in a manner that minimizes the risk of shock or electrocution to people and minimizes the possibility that such systems or equipment will start a fire.*

SECTION 1301—ELECTRICITY

[BG] 1301.1 Objective. To provide safe installation of electrical power and lighting for building systems and safe use by building occupants.

[BG] 1301.2 Functional statement. The electrical installations shall have safeguards against personal injury and the outbreak of fire.

[BG] 1301.3 Performance requirements.

[BG] 1301.3.1 Protection from live parts. People and building elements shall be protected against contact with live parts.

[BG] 1301.3.2 Isolation. The electrical installation shall allow safe isolation of devices, equipment and appliances.

[BG] 1301.3.3 Protection from excessive current. People shall be protected from the effects of current exceeding the rating of the installation.

[BG] 1301.3.4 Electromechanical stress. The installation shall protect all components and equipment from electromechanical stress caused by current exceeding its rating.

[BG] 1301.3.5 Thermal damage. Building elements shall be protected from thermal damage due to heat transfer or electric arc from electrical power installations.

[BG] 1301.3.6 Installation environment. The installation shall operate safely in the intended environment.

[BG] 1301.3.7 Flammable and explosive atmosphere. The installation shall prevent ignition of the atmosphere containing flammable or explosive elements.

[BG] 1301.3.8 Essential services and equipment. Essential services and equipment shall have a power supply protected in a manner to ensure continued operation for an appropriate time after a power failure.

[BG] 1301.3.9 Power supplier. The building electrical installation shall protect the safety features of the power supplier.

CHAPTER

14 ENERGY EFFICIENCY

User notes:

About this chapter: *Chapter 14 provides the requirements for building systems and portions of a building that impact energy use in new construction, and promotes the cost-effective use of energy.*

SECTION 1401—ENERGY EFFICIENCY

[CE] 1401.1 Objective. To facilitate efficient use of energy.

[CE] 1401.2 Functional statement. Buildings shall have provisions ensuring efficient use of nonrenewable energy.

[CE] 1401.3 Performance requirements.

[CE] 1401.3.1 Energy performance indices. To provide for the efficient use of depletable energy sources, the building envelope shall be designed and constructed within stated parameters. These parameters are called the energy performance indices. These indices are the amount of energy from a depletable energy source passing through a specified building envelope area during a specified difference in internal and external temperature. These indices are based on the region of the country as well as the use of the building. Equivalent energy performance utilizing alternative energy conservation techniques is permitted. In some cases, for certain types of buildings, the local jurisdiction has the authority to choose not to specify energy performance indices.

[CE] 1401.3.2 Temperature control. For buildings requiring a controlled temperature, the building design and construction shall take into account various factors. Normally, only insulation, types of windows and related building elements are considered when addressing energy conservation. However, to provide for the efficient use of energy, there are several other items that need to be taken into consideration, such as thermal resistance, solar radiation, air tightness and heat gain or loss from building services.

CHAPTER

15 FIRE PREVENTION AND IMPACT MANAGEMENT

User notes:

About this chapter: *There are two ways to deal with fire: it can either be prevented or managed. Chapter 15 provides requirements for fire prevention and fire management.*

Fire prevention is considered one of the more popular roles that fire codes have traditionally filled. The expectation is to limit unwanted ignition to an acceptable level. It is unreasonable to believe that all unwanted ignition can be eliminated. The fire prevention requirements of Chapter 15 are strongly linked to Chapter 16, Management of People, as many fires can be prevented by adequate training and safety procedures. Public education has a significant impact on individual awareness of fire issues and the likelihood of people exhibiting fire-safe behavior.

This chapter also provides requirements related to the management of the impact of a fire event in a building or structure. These provisions assume that a fire can occur despite any "prevent fire" measures that are being taken. The objective is to limit the impact of a fire to an acceptable level on the occupants, the general public and the building or other structure, including its contents, use and processes. Other inherent objectives include providing some level of protection for the facility's mission and revenue stream and the community tax base.

SECTION 1501—FIRE PREVENTION

Scan for Changes d48c22a

[F] 1501.1 Objective. To limit or control the likelihood that a fire will start because of the design, operation or maintenance of a facility or its systems so as to minimize impacts on people, property, processes and the environment.

[F] 1501.2 Functional statement. Facility services, systems and activities that represent a potential source of ignition or can contribute fuel to an incipient fire shall be designed, operated, managed and maintained to reduce the likelihood of a fire starting.

[F] 1501.2.1 Fuel-burning appliances and services. Fuel-burning appliances and services shall be installed in a manner that reduces their potential as sources of fire ignition.

[F] 1501.2.2 Electrical equipment, appliances and services. Electrical equipment, appliances and services shall be installed in a manner that reduces their potential as sources of fire ignition.

[F] 1501.3 Performance requirements.

[F] 1501.3.1 Ignition sources. Electrical, mechanical and chemical systems or processes and facility services capable of supplying sufficient heat under normal operating conditions or anticipated failure modes to ignite combustible system components, facility elements or nearby materials shall be designed, operated, managed and maintained to prevent the occurrence of fire.

[F] 1501.3.1.1 Uncontrolled combustion and explosion. Fuel-burning appliances and services shall be installed so that the appliance or service will not cause uncontrolled combustion or explosion.

[F] 1501.3.1.2 Fuel-burning appliances and services as sources of ignition. Fuel-burning appliances and services shall be installed so that they will not become sources of ignition.

[F] 1501.3.1.3 Sparks and arcing. Electrical equipment, appliances and services shall be installed so that they will not allow sparks or arcing to escape their enclosures.

[F] 1501.3.1.4 Electrical equipment, appliances and services. Electrical equipment, appliances and services shall be installed so that they will not become sources of ignition.

[F] 1501.3.1.5 Flammable, combustible and explosive atmospheres. Ignition sources shall be separated from areas where a flammable, combustible or explosive atmosphere may exist.

[F] 1501.3.2 Fuel sources. The quantities, configurations, characteristics or locations of combustible materials, including components or facility systems, facility elements, facility contents and accumulations of readily ignitable waste or debris shall be managed or maintained to prevent ignition by facility service equipment and other ignition sources associated with processes normally present or expected to be present within the facility.

[F] 1501.3.3 Ignition and fuel source interactions. Design, operate, and maintain facility services and facility system installation locations to prevent the occurrence or to control the extent of atmospheres likely to pose an ignition hazard.

1502—FIRE IMPACT MANAGEMENT

Scan for Changes

d35b2fb

[F] 1502.1 Objective. To provide an acceptable level of fire safety performance when facilities are subjected to fires occurring in the fire loads present in the facility during construction or alteration and throughout the intended life.

[F] 1502.2 Functional statements. Facilities shall be designed with safeguards against the spread of fire so that persons shall not suffer *serious injury* or death from a fire, and so that the magnitude of the property loss is limited and consistent with the performance groups determined in Chapter 3 as follows:

Performance Group I—High

Performance Group II—Moderate

Performance Group III—Mild

Performance Group IV—Mild

[F] 1502.2.1 Fire potential. Facilities and contents shall be maintained in a manner that limits the potential for fire.

[F] 1502.2.2 Fire impact. Facilities shall be designed, constructed and maintained to limit the fire impact to people and property.

[F] 1502.2.3 Time for evacuation. Facilities shall be designed, constructed, maintained and operated with appropriate safeguards in place to limit the spread of fire and products of combustion so that occupants have sufficient time to escape the fire.

[F] 1502.2.4 Limitation on fire spread. Facilities shall be designed, constructed, maintained and operated in such a manner that the spread of fire through a building is restricted, and that fire does not spread to adjacent properties.

[F] 1502.2.5 Wildland fires. In wildland interface areas, facilities and vegetation shall be designed, constructed, arranged and maintained in such a manner to limit the impact to the building and the facilities during a wildland fire event.

[F] 1502.2.6 Emergency responder needs. Facilities shall be arranged, constructed, maintained and operated with appropriate safeguards in place to allow firefighting personnel to perform their duties during an emergency event.

[F] 1502.2.6.1 Needs of firefighters Buildings and facilities shall be designed and constructed so that firefighters can appropriately perform rescue operations, protect property and utilize firefighting equipment and controls.

[F] 1502.2.7 Structural integrity. Facilities shall be arranged, constructed and maintained so as to limit the impact of a fire on the structural integrity of the facility.

[F] 1502.2.8 Capability of building or facility users. Facilities open to persons of varying physical and mental capabilities shall provide reasonably equivalent levels of fire safety protection for those persons to the levels it provides for persons without disabilities.

[F] 1502.3 Performance requirements. Facilities or portions thereof shall be designed, constructed and operated to normally prevent any fire from growing to a stage that would cause life loss or *serious injury*, taking into account all anticipated and permitted fire loads that would affect their performance. Facilities shall be designed to sustain local fire damage, and the facility as a whole will remain intact and not be damaged to an extent disproportionate to the original local damage.

[F] 1502.3.1 Interior surface finishes. Interior surface finishes on walls, floors, ceilings and suspended building elements shall resist the spread of fire and limit the generation of unacceptable levels of toxic gases, smoke and heat appropriate to the design performance level and associated hazards, risks and fire *safety systems* or features installed.

[F] 1502.3.2 Building materials, processes and contents. Limit quantities, configurations and combustibility of building materials, processes and contents so that fire growth and size can be controlled.

[F] 1502.3.3 Emergency responders. Where necessary, provide appropriate measures to limit fire and smoke spread and damage to acceptable levels so that firefighters are not unduly hindered in suppression or rescue operations.

[F] 1502.3.4 Detection and notification. Where human intervention or system or equipment response is necessary to limit the fire impact, provide appropriate means for detection and notification of fire.

[F] 1502.3.5 Activation of detection systems. Fire detection systems, where provided, shall activate at a fire size appropriate to the fire and life safety strategies selected.

[F] 1502.3.6 Activation of suppression systems. Automatic fire suppression systems, where provided as a means of controlling fire growth or to suppress the fire, shall deliver sufficient suppression agent to control or suppress the fire as appropriate.

[F] 1502.3.7 Control of smoke. Smoke control systems, where provided, shall limit the unacceptable spread of smoke to nonfire areas as appropriate in a manner so as not to endanger occupants.

[F] 1502.3.8 Concealed spaces. Construction in concealed spaces shall inhibit the unseen spread of fire and unacceptable movement of hot gases and smoke, appropriate to associated hazards, risks and fire *safety systems* or features installed.

[F] 1502.3.9 Vertical openings. Vertical openings shall be constructed, arranged, limited or protected to limit fire and smoke spread as appropriate to the fire and life-safety strategies selected.

[F] 1502.3.10 Wall, floor, roof and ceiling assemblies. Wall, floor, roof and ceiling assemblies forming compartments including their associated openings shall limit the spread of fire appropriate to the associated hazards, risks and fire-*safety systems* or features installed.

[F] 1502.3.11 Structural members and assemblies. Structural members and assemblies shall have a fire resistance appropriate to their function, the fire load, the predicted fire intensity and duration, the fire hazard, the height and use of the building, the proximity to other properties or structures, and any fire protection features.

[F] 1502.3.12 Exterior wall and roof assemblies' restriction of fire spread. Construction of exterior wall and roof assemblies shall restrict the spread of fire to or from adjacent buildings and from exterior fire sources, appropriate to the associated hazards, risks and fire *safety systems* or features installed.

[F] 1502.3.13 Exterior wall and roof assemblies' contribution to fire growth. Construction of exterior wall and roof assemblies shall resist the spread of fire by limiting their contribution to fire growth and development, appropriate to the associated hazards, risks and fire *safety systems* or features installed.

[F] 1502.3.14 Air handling and mechanical ventilation systems. Air handling and mechanical ventilation systems, where provided, shall be designed to avoid or limit the unacceptable spread of fire and smoke to nonfire areas as appropriate.

[F] 1502.3.15 Magnitude of fire event. Design fire events shall realistically reflect the ignition, growth and spread potential of fires and fire effluents that could occur in the fire load that may be present in the facility by its design and operational controls.

[F] 1502.3.15.1 Design fire events. Magnitudes of design fire events shall be described in terms of the potential spread of fire and fire effluents given the proposed design, arrangement, construction, furnishing and use of a building.

[F] 1502.3.15.2 Range of fire sizes. Magnitudes of design fire events shall be defined as small, medium, large and very large, based on the quantification of the design fire event as a function of the building use and associated performance group.

[F] 1502.3.15.3 Engineering analyses of potential fire scenarios. Quantification of the magnitudes of design fire events shall be based on engineering analyses of potential fire scenarios that can be expected to impact a building through its intended life. For each design fire scenario considered, the analyses shall include the ignitability of the first item, the peak heat release rate of the item first ignited, the rate of heat release and expected fire growth, and the overall fuel load, geometry, and ventilation of the space and adjoining spaces.

[F] 1502.3.15.3.1 Relationship of design fire to tolerable damage. When determining (assigning) the magnitude of a design fire event, the physical properties of the fire and its effluents shall only be considered in terms of how they impact the levels of tolerable damage. The magnitude of the fire event is not required to be characterized solely on the basis of the physical size of the fire in terms of its heat release and smoke production rates.

[F] 1502.3.15.3.2 Design parameters. Multiple design fire scenarios, ranging from small to very large design fire events, shall be considered to ensure that associated levels of tolerable damage are not exceeded as appropriate to the performance group.

[F] 1502.3.15.3.3 Factors in determining design fire scenarios. The development of design fire scenarios shall consider the use of the room of fire origin and adjoining spaces, in terms of impact on occupant, property and community welfare.

[F] 1502.3.15.3.4 Justification. Justification of the magnitudes of design fire events and design fire scenarios shall be part of the analysis prepared by the *registered design professional* and shall take into consideration the reasonableness, frequency and severity of the design fire event and design fire scenarios.

[F] 1502.3.15.3.5 Safety factors. Design fires and fire scenarios shall be chosen to provide appropriate factors of safety to provide adequate performance by accounting for the following factors:

1. Effects of uncertainties arising from construction activities.
2. Variations in the properties of materials and the characteristics of the site.
3. Accuracy limitations inherent in the methods used to predict the fire safety of the building.
4. Variations in the conditions of facilities, systems, contents and occupants.

CHAPTER

16 MANAGEMENT OF PEOPLE

User notes:

About this chapter: *Chapter 16 addresses tasks people might execute in a performance design. Many times, hazardous materials facilities depend on building occupants and users to perform certain tasks to avoid and mitigate emergencies. These activities are considered integral to the success of such facilities. Other examples include restrictions on the types of appliances allowed in the lunchroom or the number of staff available for certain types of events, such as sporting events. This chapter outlines not only the more traditional prevention and protection skills that existing prescriptive codes require, but also requires that where the actions or practices of people become a component of a design, such actions and practices must be maintained.*

SECTION 1601—MANAGEMENT OF PEOPLE

[F] 1601.1 Objective. To promote safe practices and actions of people, and to ensure that the actions and practices of people who are components of a design are maintained.

[F] 1601.2 Functional statements.

[F] 1601.2.1 Training and education for prevention of fires and other emergencies. Through training and education, ensure that people possess the necessary skills and implement the appropriate actions to prevent fires or other emergencies as appropriate to the design performance level determined in Chapter 3.

[F] 1601.2.2 Training and education for mitigation of fires and other emergencies. Through training and education, ensure that people possess the necessary skills and implement the appropriate actions during a fire or other emergency as appropriate to the design performance level determined in Chapter 3.

[F] 1601.3 Performance requirements.

[F] 1601.3.1 Identification of hazards. Provide appropriate information so that occupants and staff can assist in identifying hazards.

[F] 1601.3.2 Procedure development and training for fire or other emergency. Develop procedures and conduct training so that occupants and staff can take appropriate actions to prevent fires or other emergencies.

[F] 1601.3.3 Actions during fires or other emergencies. Provide adequate information so that occupants and staff know the appropriate actions in the event of a fire or other emergency.

[F] 1601.3.4 Procedure development and training for mitigation. Develop procedures and conduct training so that occupants and staff can take the appropriate actions in the event of a fire or other emergency.

[F] 1601.3.5 Proper handling and use of hazardous materials. Provide adequate information so that all persons involved in the handling and use of hazardous materials know the appropriate actions and safeguards for such materials.

[F] 1601.3.6 Hazardous materials emergency training. Develop procedures so that all persons involved in the handling and use of hazardous materials will take the appropriate actions in the event of an emergency.

[F] 1601.3.7 Management of procedures and training. Provide the administrative controls to ensure that the identified hazards are controlled, procedures are followed and training occurs.

[F] 1601.3.8 Validation of policies, procedures and training. Provide the administrative controls to evaluate and validate all policies, procedures and training for occupants and staff.

[F] 1601.3.9 Management of change. Whenever new occupants, staff, equipment, materials or processes are introduced, the administrative controls shall provide for appropriate education and training.

[F] 1601.3.10 Documentation of reliance on occupants and staff. Ensure that all aspects of a *performance-based design* that rely on a response or action from either occupants or staff are clearly identified and documented and that the necessary training and administrative controls are in place and maintained so that the response or action is appropriate.

CHAPTER

17 EMERGENCY ACCESS AND FACILITIES

User notes:

About this chapter: *Inevitably, various emergencies, including medical emergencies, occur at buildings and facilities. Chapter 17 deals with providing suitable access and facilities for emergency operations and responders. This chapter is intended to address the need for some manual action to preserve the safety of people, and limit damage to a building or structure and its contents.*

SECTION 1701—EMERGENCY ACCESS AND FACILITIES

Scan for Changes

fe294c9

[F] 1701.1 Objectives.

[F] 1701.1.1 Access and facilities for emergency responders. To provide and maintain means of access and facilities for emergency operations and responders.

[F] 1701.2 Functional statements. As appropriate to the design performance level in Chapter 3, the following shall be addressed:

1. Provide and maintain appropriate access for emergency vehicles.
2. Provide and maintain appropriate access for emergency responders.
3. Provide and maintain necessary staging, command and control areas, support facilities and equipment for emergency operations.
4. Provide sufficient, reliable water for firefighting operations.

[F] 1701.3 Performance requirements.

[F] 1701.3.1 Vertical and horizontal clearance for fire department apparatus. Vertical and horizontal clearance shall permit the unimpeded access of fire department apparatus inclusive of the capability for one apparatus to pass another apparatus set up and in operation.

[F] 1701.3.2 Protrusions and appurtenances from structures. Protrusions and appurtenances from structures shall not impede access, including vertical access, to the height of fire department aerial apparatus.

[F] 1701.3.3 Surfaces for fire department apparatus. Fire department access shall be on surfaces permitting year-round, all-weather travel at a grade appropriate for the fire apparatus.

[F] 1701.3.4 Hose length limitations. Access to structures shall afford the fire department the ability to deploy and operate hose lines without the need to extend the standard hose line utilized by the fire department having jurisdiction.

[F] 1701.3.5 Control valve locations. Within structures, means for the deployment and operation of hose lines by emergency responders shall be provided such that control valves for lines shall be no further from potential fire sources than the length of hose packs employed by a single engine company of the fire department having jurisdiction.

[F] 1701.3.6 Water supply. Water supply for fire department operations shall be from a reliable source with ready *access* acceptable to the fire department and capable of supporting firefighting operations.

[F] 1701.3.7 Horizontal or vertical conveyance. Means of horizontal or vertical conveyance shall be provided where necessary to support firefighting and emergency support functions.

[F] 1701.3.8 Staging areas. Where interior operations may be necessary, areas to stage equipment and from which to safely conduct and to control suppression operations shall be provided.

[F] 1701.3.9 Interaction of access and means of egress. Exterior and interior egress and emergency access shall be arranged and maintained so that building occupants and emergency responders are unimpeded as each accomplishes its objectives of egress of occupants and access by emergency responders.

[F] 1701.3.10 Interior and exterior staging. Where necessary to ensure timely and effective emergency operations, interior or exterior areas shall be provided for the staging of equipment and apparatus.

[F] 1701.3.11 On-site equipment. Where necessary to ensure timely and effective emergency operations, firefighting equipment or other equipment to support such operations shall be provided and maintained readily available for use by emergency responders.

CHAPTER

18 EMERGENCY RESPONDER SAFETY

User notes:

About this chapter: *Chapter 18 addresses the major issues impacting emergency responder safety. The prescriptive codes have always made provisions to lessen the dangers to emergency responders, firefighters in particular; however, this code takes the next step and places these concerns into a separate chapter. This does not mean that provisions found elsewhere in this document do not also pertain to firefighter safety, but only that issues specific thereto are addressed in this chapter. Such provisions can also be found in the chapters on egress, access and hazardous materials.*

SECTION 1801—EMERGENCY RESPONDER SAFETY

[F] 1801.1 Objective. To protect emergency responders from unreasonable risks during emergencies.

[F] 1801.2 Functional statements. As appropriate to the design performance level determined in Chapter 3, the following shall be provided:

1. Information to responders regarding hazards present at the building or premises.
2. Protection against unanticipated structural collapse.
3. Appropriate fire service communications capability.

[F] 1801.3 Performance requirements.

[F] 1801.3.1 Identification of hazards. Where hazards are present in the building, facility or premises that could endanger emergency responders beyond what would normally be anticipated, means shall be provided to alert the responders to the hazards.

[F] 1801.3.2 Signage. Signage shall be provided as needed to identify special hazards to the emergency responders (and to the degree applicable, the nature of the hazard).

[F] 1801.3.3 Collapse. Buildings and structures shall be designed, constructed, loaded and maintained so that the potential for structural collapse is predictable based on the construction method, building condition and fire size, location and duration.

[F] 1801.3.4 Communication systems. Communication systems for use by the emergency responders must be provided where the size, construction or complexity of the building cause the emergency responders' communication methods to be ineffective or unreliable.

CHAPTER

19 HAZARDOUS MATERIALS

User notes:

About this chapter: *This chapter and Chapter 50 of the International Fire Code® have similar aims: to protect occupants of the building, people in the surrounding area, emergency response personnel and property from acute consequences associated with unintended or unauthorized releases of hazardous materials. Like the prescriptive International Fire Code® and International Building Code®, this code encourages the use of both accident prevention and control measures to reduce risk. The provisions of this chapter are also found within the Section 5001.3 of the* International Fire Code *and offered as a performance-based alternative.*

SECTION 1901—HAZARDOUS MATERIALS

Scan for Changes

6dc24e4

[F] 1901.1 Objective. To protect people and property from the consequences of unauthorized discharge, fires or explosions involving hazardous materials.

[F] 1901.2 Functional statements.

[F] 1901.2.1 Prevention. Provide adequate safeguards to minimize the risk of unwanted releases, fires or explosions involving hazardous materials as appropriate to the design performance level determined in Chapter 3.

[F] 1901.2.2 Mitigation. Provide adequate safeguards to minimize the consequences of an unsafe condition involving hazardous materials during normal operations and in the event of an abnormal condition in accordance with the design performance level determined in Chapter 3.

[F] 1901.3 Performance requirements.

[F] 1901.3.1 Properties of hazardous materials. The properties of hazardous materials on site shall be known and shall be available to employees, neighbors and code enforcement officials.

[F] 1901.3.2 Reliability of equipment and operations. Equipment and operations involving hazardous materials shall be designed, installed and maintained to ensure that they reliably operate as intended.

[F] 1901.3.3 Prevention of unintentional reaction or release. Adequate safeguards shall be provided to minimize the risk of an unintentional reaction or release that could endanger people or property.

[F] 1901.3.4 Spill mitigation. Spill containment systems or means to render a spill harmless to people or property shall be provided where a spill is determined to be a plausible event and where such an event would endanger people or property not in the immediate area of the spill.

[F] 1901.3.5 Ignition hazards. Adequate safeguards shall be provided to minimize the risk of exposing combustible hazardous materials to unintended sources of ignition.

[F] 1901.3.6 Protection of hazardous materials. Adequate safeguards shall be provided to minimize the risk of exposing hazardous materials to a fire or physical damage whereby such exposure could endanger or lead to the endangerment of people or property.

[F] 1901.3.7 Exposure hazards. Adequate safeguards shall be provided to minimize the risk of and limit damage from a fire or explosion involving explosive hazardous materials whereby such fire or explosion could endanger or lead to the endangerment of people or property.

[F] 1901.3.8 Detection of gas or vapor release. Where a release of hazardous materials gas or vapor would cause immediate harm to persons or property and where such materials would not be detectable at the danger threshold by sight or smell, an adequate means of detecting, diluting or otherwise mitigating the dangerous effects of a release shall be provided.

[F] 1901.3.9 Reliable power source. Where a power supply is relied on to prevent or control an emergency condition that could endanger people or property, the power supply shall be from a reliable source.

[F] 1901.3.10 Ventilation. Where ventilation is necessary to limit the risk of creating an emergency condition resulting from normal or abnormal operations, an adequate means of ventilation shall be provided.

[F] 1901.3.11 Process hazard analyses. Process hazard analyses shall be conducted as necessary to reasonably ensure protection of people and property from dangerous conditions involving hazardous materials.

[F] 1901.3.12 Written procedures and enforcement for prestartup safety review. Written documentation of prestartup safety review procedures shall be developed and enforced to ensure that operations are initiated in a safe manner. The process of developing and updating such procedures shall involve participation of affected employees.

[F] 1901.3.13 Written procedures and enforcement for operation and emergency shutdown. Written documentation of operating procedures and procedures for emergency shutdown shall be developed and enforced to ensure that operations are conducted in a safe manner. The process of developing and updating such procedures shall involve participation of affected employees.

[F] 1901.3.14 Written procedures and enforcement for management of change. A written plan for management of change shall be developed and enforced. The process of developing and updating the plan shall involve participation of affected employees.

[F] 1901.3.15 Written procedures for action in the event of emergency. A written emergency response plan shall be developed to ensure that proper actions are taken in the event of an emergency, and the plan shall be followed if an emergency condition occurs. The process of developing and updating the plan shall involve participation of affected employees.

[F] 1901.3.16 Written procedures for investigation and documentation of accidents. Written procedures for investigation and documentation of accidents shall be developed, and accidents shall be investigated and documented in accordance with these procedures.

[F] 1901.3.17 Consequence analysis. Where an accidental release of hazardous materials could endanger people or property off site, an analysis of the expected consequences of a plausible release shall be performed and utilized in the analysis and selection of active and passive hazard mitigation controls.

[F] 1901.3.18 Safety audits. Safety audits shall be conducted on a periodic basis to verify compliance with the requirements of this chapter.

[F] 1901.3.19 Levels of impact or damage. Levels of impact or damage shall comply with the requirements of Section 304 for design performance levels.

[F] 1901.3.19.1 General. Magnitudes of design events shall reflect the ignition, spill or release, growth and spread potential of hazardous materials that can be reasonably expected to impact buildings and facilities as designed or constructed.

[F] 1901.3.19.2 Design hazardous materials release or reaction events. Magnitudes of design events are described in terms of the potential effects given the proposed design, arrangement, construction, furnishing and use of a building or facility.

[F] 1901.3.19.3 Range of event sizes. Magnitudes of design events shall be defined as small, medium, large and very large, where the quantification of the design event is a function of building or facility use and associated performance group.

[F] 1901.3.19.4 Engineering analysis of potential event scenarios. Quantification of the magnitudes of design events shall be based on engineering analyses of potential scenarios that can be expected to impact a building or facility through its intended life. For each design scenario considered, the analyses shall include the ignitability, reactivity, spill or release potential, the peak release rate, the rate of continued release and expected incident growth, the overall quantity, the toxicity, chemical state and other specific hazards of the material and its impacts on people and property. The physical characteristics and ventilation of the space or area and adjoining spaces or areas shall be considered.

[F] 1901.3.19.5 Design parameters. Multiple scenarios, ranging from small to very large design events, must be considered to ensure that associated levels of tolerable damage are not exceeded as appropriate to the performance group.

[F] 1901.3.19.6 Factors in determining design events scenarios. The use of the room or area of event and adjoining spaces, in terms of occupant risk, property protection and community welfare (importance) factors, shall be considered in the development of design scenarios.

[F] 1901.3.19.7 Justification. Justification of the magnitudes of design events shall be part of the analysis prepared by the *registered design professional*.

CHAPTER

20 REFERENCED STANDARDS

User note:

About this chapter: *Chapter 20 notes the standard that is referenced in this code. This chapter provides the location where future standards referenced by this code can be listed. The standard is part of the code to the extent of the reference to the standard. Compliance with the referenced standard is necessary for compliance with this code. By providing this specifically adopted standard, the construction and installation requirements necessary for compliance with the code can be readily determined. The basis for code compliance is, therefore, established and available on an equal basis to the building code official, contractor, designer and owner.*

This standard is listed herein by the promulgating agency of the standard, the standard identification, the effective date and title, and the sections of this document that reference the standard.

ASCE/SEI *American Society of Civil Engineers*

1801 Alexander Bell Drive, Reston, VA 20191-4400

7—2022: Minimum Design Loads and Associated Criteria for Buildings and Other Structures

501.3.4, 501.3.5

APPENDIX A
RISK FACTORS OF USE AND OCCUPANCY CLASSIFICATIONS

User notes:

About this appendix: *Appendix A classifies buildings, structures and portions thereof by their primary use in order to facilitate design and construction in accordance with other provisions of this code. When determining the design performance level, the building or structure needs to be assigned to a performance group. This appendix provides guidance as to the use group or occupancy classification of the building to determine the performance group when applying Table 303.1.*

Code development reminder: *Code change proposals to this appendix will be considered by the IBC—General Code Development Committee during the 2025 (Group B) Code Development Cycle.*

SECTION A101—OBJECTIVE

[BG] A101.1 Objective. To identify the primary uses of buildings, structures and portions of buildings and structures; to identify risk factors associated with these uses; and to facilitate design and construction in accordance with other provisions of this code. These preliminary assumptions must be documented and verified as valid in any particular case.

SECTION A102—FUNCTIONAL STATEMENTS

[BG] A102.1 Functional statements. In determining the primary use and occupancy classification of a building, structure or portion of a building or structure, the following shall be considered.

[BG] A102.1.1 Functions. The principal purpose or function of the building or structure.

[BG] A102.1.2 Risks. The hazard-related risk(s) to the users.

1. In determining the principal purpose or function of the building or structure, the use and occupancy classifications found in the *International Building Code* shall be used.
2. In determining the hazard-related risk(s) to users of buildings and structures, the following shall be considered:
   - 2.1. The nature of the hazard, whether it is likely to originate internal or external to the structure, and how it may impact the occupants, the structure and the contents.
   - 2.2. The number of persons normally occupying, visiting, employed in or otherwise using the building, structure or portion of the building or structure.
   - 2.3. The length of time the building is normally occupied by people.
   - 2.4. Whether people normally sleep in the building.
   - 2.5. Whether the building occupants and other users are expected to be familiar with the building layout and means of egress.
   - 2.6. Whether a significant percentage of the building occupants are, or are expected to be, members of vulnerable population groups such as infants, young children, elderly persons, persons with physical disabilities, persons with mental disabilities or persons with other conditions or impairments that could affect their ability to make decisions, egress without the physical assistance of others or tolerate adverse conditions.
   - 2.7. Whether a significant percentage of building occupants and other users have familiar or dependent relationships.

SECTION A103—USE AND OCCUPANCY CLASSIFICATION

[BG] A103.1 General. Buildings shall be classified in accordance with the *International Building Code* and as modified by applying the risk considerations in Section A102.1.2.

Scan for Changes
a0b1cc0

[BG] A103.1.1 Assembly. A building, structure or portion of a building or structure in which persons gather for purposes such as civic, social or religious functions, recreation, food and drink consumption, or awaiting transportation. Unless otherwise modified under a specific sub-use classification, occupants, visitors and employees shall be assumed to be awake, alert, predominantly able to exit without the assistance of others, and unfamiliar with the building or structure. Vulnerable populations of many types may be expected to be present; however, the buildings are normally occupied for only short periods of time. It shall be assumed that:

1. Risks of injury and health assumed by occupants and visitors during their use of the building or structure are predominantly involuntary.
2. Public expectations regarding the protection afforded those occupying, visiting or working in an assembly building, structure or portion thereof are high.

[BG] A103.1.1.1 A-1. Assembly uses, usually with fixed seating, intended for the production and viewing of the performing arts or motion pictures.

[BG] A103.1.1.2 A-2. Assembly uses intended for food and drink consumption. It may be assumed that some portion of the population within this use group will be consuming alcohol, that loud and distracting sounds will be present, and that flashing lights may be present under normal conditions.

[BG] A103.1.1.3 A-3. Assembly uses intended for worship, recreation or amusement, and other assembly uses not classified elsewhere in Use Group A.

[BG] A103.1.1.4 A-4. Assembly uses intended for viewing of indoor sporting events and activities with spectator seating. It may be assumed that some portion of the population within this use group will be consuming alcohol, that loud and distracting sounds will be present and that flashing lights may be present under normal conditions.

[BG] A103.1.1.5 A-5. Assembly uses intended for participation in or viewing of outdoor activities. It may be assumed that some portion of the population within this use group will be consuming alcohol, that loud and distracting sounds will be present and that flashing lights may be present under normal conditions.

[BG] A103.1.2 Business. A building, structure or portion of a building or structure for office, professional or service-type transactions, including storage of records and accounts. It shall be assumed that:

1. Occupants, visitors and employees are awake, alert, predominantly able to exit without the assistance of others and familiar with the building or structure.
2. Risks of injury and health assumed by occupants, visitors and employees during their use of the building or structure are predominantly involuntary.
3. Public expectations regarding the protection afforded those occupying, visiting or working in a business building, structure or portion thereof are neither unusually high nor unusually low.

[BG] A103.1.3 Educational. A building, structure or portion of a building or structure in which six or more persons, generally under the age of 18, gather for formal educational purposes. It shall be assumed that:

1. Occupants, visitors and employees are awake, alert and familiar with the building or structure.
2. Persons under the age of 10 will require assistance in exiting, and that persons 10 years of age and older will predominantly be able to exit without assistance.
3. Risks of injury and health assumed by occupants, visitors and employees during their use of the building or structure are predominantly involuntary.
4. Public expectations regarding the protection afforded those occupying, visiting or working in an educational building, structure or portion thereof are high.

[BG] A103.1.4 Factory-industrial. A building, structure or portion of a building or structure that involves assembling, disassembling, fabricating, finishing, manufacturing, packaging, repair or processing operations that are not classified as hazardous. Unless otherwise modified under a specific sub-use classification, occupants, visitors and employees shall be assumed to be awake, alert, predominantly able to exit without the assistance of others, and familiar with the building or structure. It shall be assumed that:

1. Risks of injury and health assumed by occupants, visitors and employees during their use of the building or structure are predominantly voluntary.
2. Public expectations regarding the protection afforded those occupying, visiting or working in a factory-industrial building, structure or portion thereof are neither unusually high nor unusually low.

[BG] A103.1.4.1 F-1, Moderate hazard. Factory-industrial uses that are not classified as F-2, low hazard.

[BG] A103.1.4.2 F-2, Low hazard. Factory-industrial uses that involve the fabrication or manufacturing of noncombustible materials, which during finishing, packing or processing, do not involve a significant fire hazard.

[BG] A103.1.5 Hazardous. A building, structure or portion of a building or structure that involves the manufacturing, processing, generation or storage of corrosive, highly toxic, highly combustible, flammable or explosive materials that constitute a high fire or explosion hazard, including loose combustible fibers, dust and unstable materials. Unless otherwise modified under a specific subuse classification, occupants, visitors and employees shall be assumed to be awake, alert, predominantly able to exit without the assistance of others and familiar with the building or structure. It shall be assumed that:

1. Risks of injury and health assumed by occupants, visitors and employees during their use of the building or structure are predominantly voluntary.
2. The occupants, visitors and employees have little control over hazards imposed on them during their use of the building or structure.
3. Public expectations regarding the protection afforded those occupying, visiting or working in a hazardous building, structure or portion thereof are high.

[BG] A103.1.5.1 H-1, Detonation hazard. A building or structure that contains materials that present a detonation hazard.

[BG] A103.1.5.2 H-2, Deflagration hazard. A building or structure that contains materials that present a deflagration hazard or a hazard of accelerated burning.

[BG] A103.1.5.3 H-3, Combustion or physical hazard. A building or structure that contains significant quantities of materials that readily support combustion or that present a significant physical hazard.

[BG] A103.1.5.4 H-4, Health hazard. A building or structure that contains significant quantities of materials that are health hazards.

[BG] A103.1.5.5 H-5, Production material hazard. A semiconductor fabrication facility or comparable research and development area in which hazardous production materials (HPM) are used and the aggregate quantity poses a significant hazard.

[BG] A103.1.6 Institutional. A building, structure or portion of a building or structure in which persons having physical or mental limitations because of health or age are harbored for medical treatment or other care or treatment, or in which people are detained for penal or correctional purposes, or in which the liberty of the occupants is restricted. Unless otherwise modified under a specific subuse classification, occupants and visitors shall not be assumed to be awake, alert, able to exit without the assistance of others, or be familiar with the building or structure. Employees shall be assumed to be awake, alert, predominantly able to exit without the assistance of others, and familiar with the building or structure. It shall be assumed that:

1. Risks of injury and health assumed by occupants and visitors during their use of the building or structure are predominantly involuntary.
2. Risks of injury and health assumed by employees during their use of the building or structure are predominantly voluntary.
3. Public expectations regarding the protection afforded those occupying, visiting or working in an institutional building, structure or portion thereof are moderate to high.

[BG] A103.1.6.1 I-1. A building, structure or portion of a building or structure housing more than 16 persons on a 24-hour basis who, because of age, mental disability or other reasons, live in a supervised residential environment where personal care services are provided. It shall be assumed that:

1. The occupants are mostly capable of responding to an emergency situation without physical assistance from staff.
2. Risk of injury and risk to health assumed by occupants and visitors during their use of the building or structure are nominally moderate.
3. Public expectations regarding the protection afforded those occupying, visiting or working in an I-1 institutional building, structure or portion thereof are high.

[BG] A103.1.6.2 I-2. A building, structure or portion of a building or structure used for medical, surgical, psychiatric, nursing or custodial care on a 24-hour basis for more than five persons who are not capable of self-preservation. It shall be assumed that:

1. The occupants are incapable of responding to an emergency situation without physical assistance from staff.
2. Risk of injury and risk to health assumed by occupants and visitors during their use of the building or structure are nominally high.
3. Public expectations regarding the protection afforded those occupying, visiting or working in an I-2 institutional building, structure or portion thereof are very high.

[BG] A103.1.6.3 I-3. A building or structure that is inhabited by more than five persons who are under some restraint security.

[BG] A103.1.6.3.1 I-3.1. A building in that free movement is allowed from sleeping areas and other areas where access or occupancy is permitted to the exterior via a means of egress without restraint. It shall be assumed that:

1. The occupants are mostly incapable of responding to an emergency situation without physical assistance from staff.
2. Risk of injury and risk to health assumed by occupants and visitors during their use of the building or structure are nominally low.
3. Public expectations regarding the protection afforded those occupying, visiting or working in an I-3.1 institutional building, structure or portion thereof are neither unusually high nor unusually low.

[BG] A103.1.6.3.2 I-3.2. A building in which free movement is allowed from sleeping areas and any other occupied smoke compartment to one or more other smoke compartments. Egress to the exterior is impeded by locked exits. It shall be assumed that:

1. The occupants are incapable of responding to an emergency situation without physical assistance from staff (for example, door release).
2. Risk of injury and risk to health assumed by occupants and visitors during their use of the building or structure are nominally moderate.
3. Public expectations regarding the protection afforded those occupying, visiting or working in an I-3.2 institutional building, structure or portion thereof are neither unusually high nor unusually low.

[BG] A103.1.6.3.3 I-3.3. A building in which free movement is allowed within individual smoke compartments, and where egress is impeded by remote-controlled release of means of egress from one smoke compartment to another smoke compartment. It shall be assumed that:

1. The occupants are incapable of responding to an emergency situation without physical assistance from staff (for example, door release).
2. Risk of injury and risk to health assumed by occupants and visitors during their use of the building or structure are nominally moderate.
3. Public expectations regarding the protection afforded those occupying, visiting or working in an I-3.3 institutional building, structure or portion thereof are neither unusually high nor unusually low.

[BG] A103.1.6.3.4 I-3.4. A building in which free movement is restricted from an occupied space. Remote-controlled release is provided to permit movement from sleeping rooms, activity spaces and other occupied areas within the smoke compartment to other smoke compartments. It shall be assumed that:

1. The occupants are incapable of responding to an emergency situation without physical assistance from staff (for example, door release).
2. Risk of injury and risk to health assumed by occupants and visitors during their use of the building or structure are nominally high.
3. Public expectations regarding the protection afforded those occupying, visiting or working in an I-3.4 institutional building, structure or portion thereof are neither unusually high nor unusually low.

[BG] A103.1.6.3.5 I-3.5. Buildings in which free movement is restricted from an occupied space. Staff-controlled release is provided to permit movement from sleeping rooms, activity spaces and other occupied areas within the smoke compartment to other smoke compartments. It shall be assumed that:

1. The occupants are incapable of responding to an emergency situation without physical assistance from staff (for example, door release).
2. Risk of injury and risk to health assumed by occupants and visitors during their use of the building or structure are nominally high.
3. Public expectations regarding the protection afforded those occupying, visiting or working in an I-3.5 institutional building, structure or portion thereof are neither unusually high nor unusually low.

[BG] A103.1.6.4 I-4. A building or structure occupied by persons of any age who receive custodial care for less than 24 hours by individuals other than parents or guardians, relatives by blood, marriage or adoption, and in a place other than the home of the person receiving care.

[BG] A103.1.6.4.1 I-4.1, Adult-care facilities. A facility that provides accommodation for more than five unrelated adults and provides supervision and personal care services. It shall be assumed that:

1. The occupants are mostly capable of responding to an emergency situation without physical assistance from staff.
2. Risk of injury and risk to health assumed by occupants and visitors during their use of the building or structure are nominally moderate.
3. Public expectations regarding the protection afforded those occupying, visiting or working in an I-4.1 institutional building, structure or portion thereof are high.

[BG] A103.1.6.4.2 I-4.2, Child-care facilities. A facility that provides accommodation for more than five children, $2^1/_2$ years of age or less. It shall be assumed that:

1. The occupants are incapable of responding to an emergency situation without physical assistance from staff.
2. Risk of injury and risk to health assumed by occupants and visitors during their use of the building or structure are nominally high.
3. Public expectations regarding the protection afforded those occupying, visiting or working in an I-4.2 institutional building, structure or portion thereof are very high.

[BG] A103.1.7 Mercantile. A building, structure or portion of a building or structure for the display and sale of merchandise that involves stocks of goods, wares or merchandise incidental to such purposes and available to the public. It shall be assumed that:

1. Occupants, visitors and employees are awake, alert, predominantly able to exit without the assistance of others and familiar with the building or structure.
2. Risk of injury and risk to health assumed by occupants, visitors and employees during their use of the building or structure are predominantly involuntary and low.
3. Public expectations regarding the protection afforded those occupying, visiting or working in a mercantile building, structure or portion thereof are neither unusually high nor unusually low.

[BG] A103.1.8 Residential. A building, structure or portion of a building or structure for sleeping accommodations where not classified as institutional.

[BG] A103.1.8.1 R-1, Transient. A residential occupancy where occupants are primarily transient in nature (staying less than 30 days).

[BG] A103.1.8.1.1 R-1.1, Hotel/motel. It shall be assumed that:

1. Occupants and visitors are not awake, alert, able to exit without the assistance of others or familiar with the building or structure.
2. Employees are awake, alert, predominantly able to exit without the assistance of others and familiar with the building or structure.
3. Risks of injury and health assumed by occupants and visitors during their use of the building or structure are predominantly involuntary.
4. Risk of injury and risk to health assumed by employees during their use of the building or structure are predominantly voluntary and low.

5. Public expectations regarding the protection afforded those occupying, visiting or working in the R-1.1 residential building, structure or portion thereof are high.

[BG] A103.1.8.1.2 R-1.2, Boarding houses. It shall be assumed that:

1. Occupants and visitors are not awake, alert, able to exit without the assistance of others or familiar with the building or structure.
2. Employees are awake, alert, predominantly able to exit without the assistance of others and familiar with the building or structure.
3. Risk of injury and risk to health assumed by occupants and visitors during their use of the building or structure are predominantly involuntary.
4. Risk of injury and risk to health assumed by employees during their use of the building or structure are predominantly voluntary and moderate.
5. Public expectations regarding the protection afforded those occupying, visiting or working in the R-1.2 residential building, structure or portion thereof are moderate.

[BG] A103.1.8.2 R-2, Multitenant residential. A residential occupancy where the occupants are primarily permanent in nature and that contains more than two dwelling units. It shall be assumed that:

1. Occupants and visitors are not awake, alert or able to exit without the assistance of others.
2. Occupants and visitors are familiar with the building or structure.
3. Risk of injury and risk to health assumed by occupants and visitors during their use of the building or structure are predominantly voluntary.
4. Public expectations regarding the protection afforded those occupying, visiting or working in the R-2 residential building, structure or portion thereof are neither unusually high nor unusually low.

[BG] A103.1.8.3 R-3, One- and two-family residential. A residential occupancy where the occupants are primarily permanent in nature, not classified as R-1 or R-2, and that does not contain more than two dwelling units. It shall be assumed that:

1. Occupants and visitors are not awake, alert or able to exit without the assistance of others.
2. Occupants and visitors are familiar with the building or structure.
3. Risk of injury and risk to health assumed by occupants and visitors during their use of the building or structure are predominantly voluntary.
4. Public expectations regarding the protection afforded those occupying, visiting or working in the R-3 residential building, structure or portion thereof are neither unusually high nor unusually low.

[BG] A103.1.8.4 R-4, Residential care. A residential occupancy that includes buildings arranged for occupancy as residential care/assisted living facilities including not more than 16 occupants excluding staff. It shall be assumed that:

1. Occupants and visitors are not awake, alert, able to exit without the assistance of others or familiar with the building or structure.
2. Employees are awake, alert, predominantly able to exit without the assistance of others and familiar with the building or structure.
3. Risk of injury and risk to health assumed by occupants and visitors during their use of the building or structure are predominantly involuntary.
4. Risk of injury and risk to health assumed by employees during their use of the building or structure are predominantly voluntary and moderate.
5. Public expectations regarding the protection afforded those occupying, visiting or working in the R-4 residential building, structure or portion thereof are high.

[BG] A103.1.9 Special use. A building or structure that may not be fully described or considered under the other use group classifications or for which unique or special consideration shall be given.

[BG] A103.1.9.1 SP-1, Covered mall building. A building or structure not exceeding three floor levels in height, enclosing a number of occupancies and tenancies use groups, wherein two or more tenants have a main entrance into one or more roofed or covered common pedestrian areas shared by the tenants. It shall be assumed that:

1. Occupants, visitors and employees are awake, alert, predominantly able to exit without the assistance of others and familiar with the building or structure.
2. Risk of injury and risk to health assumed by occupants, visitors and employees during their use of the building or structure are predominantly involuntary.
3. Public expectations regarding the protection afforded those occupying, visiting or working in such a building, structure or portion thereof are neither unusually high nor unusually low.

[BG] A103.1.9.2 SP-2, High-rise building. A building or structure having occupied floors located more than 75 feet (22 860 mm) above the lowest level of fire department vehicle access that contain any use group classification or combination of use group classifications. The assumed risk levels, hazard levels and occupant characteristics shall be appropriate to the uses present within the building, and the structural, fire protection and means-of-egress features shall be designed to accommodate the highest risk level present in the building. It shall be assumed that public expectations regarding the protection afforded those occupying, visiting or working in such a building, structure or portion thereof are high.

[BG] A103.1.9.3 SP-3, Atrium. An opening through two or more floor levels of a building defined by other use groups, other than for enclosed stairways, elevators, hoistways, escalators, plumbing, electrical, air-conditioning or other equipment, that is closed at the top and not defined as a mall. Risk and hazard levels shall correspond to the use group of the building within which the atrium is located.

[BG] A103.1.9.4 SP-4, Underground building. Building spaces having a floor level used for human occupancy more than 30 feet (9140 meters) or more than one story below the level of exit discharge. Risk and hazard levels shall correspond to the use group of the building.

[BG] A103.1.9.5 SP-5, Mechanical-access open parking garage. A structure that is used exclusively for the parking or storage of private motor vehicles, where for natural ventilation purposes, the exterior of the structure has uniformly distributed openings of not less than 20 percent of the total perimeter wall area of each tier on two or more sides; that employs parking machines, lifts, elevators or other mechanical devices for vehicles moving from and to street level; and in which public occupancy is prohibited above street level. It shall be assumed that:

1. Occupants, visitors and employees are awake, alert, predominantly able to exit without the assistance of others and unfamiliar with the building or structure.
2. Risk of injury and risk to health assumed by occupants, visitors and employees during their use of the building or structure are predominantly involuntary and low.
3. Public expectations regarding the protection afforded those occupying, visiting or working in such a building, structure or portion thereof are neither unusually high nor unusually low.

[BG] A103.1.9.6 SP-6, Ramp-access open parking garage. A structure that is used exclusively for the parking or storage of private motor vehicles, where for natural ventilation purposes, the exterior of the structure has uniformly distributed openings of not less than 20 percent of the total perimeter wall area of each tier on two or more sides; that employs a series of interconnecting ramps between tiers permitting the movement of vehicles under their own power from and to street level. It shall be assumed that:

1. Occupants, visitors and employees are awake, alert, predominantly able to exit without the assistance of others and unfamiliar with the building or structure.
2. Risk of injury and risk to health assumed by occupants, visitors and employees during their use of the building or structure are predominantly involuntary and low.
3. Public expectations regarding the protection afforded those occupying, visiting or working in such a building, structure or portion thereof are neither unusually high nor unusually low.

[BG] A103.1.9.7 SP-7, Enclosed parking garage. A structure used exclusively for the parking or storage of private motor vehicles that does not meet the requirements of SP-5 or SP-6. It shall be assumed that:

1. Occupants, visitors and employees are awake, alert, predominantly able to exit without the assistance of others and unfamiliar with the building or structure.
2. Risk of injury and risk to health assumed by occupants, visitors and employees during their use of the building or structure are predominantly involuntary and low.
3. Public expectations regarding the protection afforded those occupying, visiting or working in such a building, structure or portion thereof are neither unusually high nor unusually low.

[BG] A103.1.9.8 SP-8, Motor vehicle service station. A facility containing storage tanks, pumps and attendant facilities for the purpose of fueling gasoline- and diesel-powered motor vehicles; and a building, structure or portion of a building or structure that is used for changing tires, oil, filters or other minor repair of motor vehicles, and for motor vehicle safety and emissions inspections. It shall be assumed that:

1. Occupants, visitors and employees are awake, alert, predominantly able to exit without the assistance of others and familiar with the building or structure.
2. Risk of injury and risk to health assumed by occupants, visitors and employees during their use of the building or structure are predominantly involuntary and low.
3. Public expectations regarding the protection afforded those occupying, visiting or working in such a building, structure or portion thereof are neither unusually high nor unusually low.

[BG] A103.1.9.9 SP-9, Motor vehicle repair garage. A building, structure or portion of a building or structure that is used for painting, body and fender work, engine overhauling or other major repair of motor vehicles. It shall be assumed that:

1. Occupants, visitors and employees are awake, alert, predominantly able to exit without the assistance of others and familiar with the building or structure.
2. Risk of injury and risk to health assumed by occupants, visitors and employees during their use of the building or structure are predominantly involuntary and low.
3. Public expectations regarding the protection afforded those occupying, visiting or working in such a building, structure or portion thereof are neither unusually high nor unusually low.

[BG] A103.1.9.10 SP-10, Motion picture projection room. A room in which ribbon-type cellulose, acetate or other safety film is utilized in conjunction with electric arc, xenon or other light-source projection equipment that, when operated, may result in the production of hazardous gases, dust or radiation. It shall be assumed that:

1. Visitors and employees are awake, alert and predominantly able to exit without the assistance of others.
2. Risk of injury and risk to health assumed by occupants, visitors and employees during their use of the building or structure are predominantly involuntary and low.
3. Public expectations regarding the protection afforded those occupying, visiting or working in such a building, structure or portion thereof are neither unusually high nor unusually low.

[BG] A103.1.9.11 SP-11, Stages and platforms. Spaces within buildings and structures, often raised above floor level, used for entertainment, presentations and similar purposes. It shall be assumed that:

1. Users are awake, alert and predominantly able to exit without the assistance of others.
2. Risk of injury and risk to health assumed by occupants, visitors and employees during their use of the building or structure are predominantly involuntary and low.
3. Public expectations regarding the protection afforded those occupying, visiting or working in such a building, structure or portion thereof are neither unusually high nor unusually low.

[BG] A103.1.9.12 SP-12, Special amusement area. A temporary or permanent building or portion thereof that is occupied for amusement, entertainment or educational purposes and is arranged in a manner that meets one or more of the following descriptions:

1. Makes the means of egress path not readily apparent due to visual or audio distractions.
2. Intentionally confounds identification of the means of egress path.
3. Otherwise makes the means of egress path not readily available because of the nature of the attraction or mode of conveyance through the building or structure.

It shall be assumed that:

1. Occupants, visitors and employees are awake, alert, predominantly able to exit without the assistance of others and unfamiliar with the area, building or structure.
2. Risk of injury and risk to health assumed by occupants, visitors and employees during their use of the area, building or structure are predominantly involuntary and high.
3. Public expectations regarding the protection afforded those occupying, visiting or working in such an area, building or structure are high.

[BG] A103.1.9.13 SP-13, Aircraft-related structure. A building, structure or portion of a building or structure used for air traffic control, aircraft storage and maintenance, or helicopter landing and fueling. It shall be assumed that:

1. Occupants, visitors and employees are awake, alert, predominantly able to exit without the assistance of others and familiar with the building or structure.
2. Risk of injury and risk to health assumed by occupants, visitors and employees during their use of the building or structure are predominantly involuntary and low.
3. Public expectations regarding the protection afforded those occupying, visiting or working in such a building, structure or portion thereof are neither unusually high nor unusually low.

[BG] A103.1.10 Storage. A building, structure or portion of a building or structure for storage that is not classified as hazardous. Unless otherwise modified under a specific sub-use classification, occupants, visitors and employees shall be assumed to be awake, alert, predominantly able to exit without the assistance of others and familiar with the building or structure. It shall be assumed that:

1. Risk of injury and risk to health assumed by occupants, visitors and employees during their use of the building or structure are predominantly voluntary.
2. Public expectations regarding the protection afforded those occupying, visiting or working in such a building, structure or portion thereof are neither unusually high nor unusually low.

[BG] A103.1.10.1 ST-1, Moderate hazard. A building or structure occupied for storage uses, which may contain materials that present moderate fire, explosion, corrosive, toxic or health hazards.

[BG] A103.1.10.2 ST-2, Low hazard. A building or structure occupied for storage uses that does not contain materials that present significant fire, explosion, corrosive, toxic or health hazards.

[BG] A103.1.11 Utility and miscellaneous. A building, structure or portion of a building or structure that is accessory to one- and two-family residential buildings, is used to house livestock and livestock feed or is not covered by other provisions of this section. It shall be assumed that:

1. Occupants, visitors and employees are awake, alert, predominantly able to exit without the assistance of others and familiar with the building or structure.
2. Risk of injury and risk to health assumed by occupants, visitors and employees during their use of the building or structure are predominantly voluntary.
3. Public expectations regarding the protection afforded those occupying, visiting or working in such a building, structure or portion thereof are neither unusually high nor unusually low.

APPENDIX

B

WORKSHEET FOR ASSIGNING SPECIFIC STRUCTURES TO PERFORMANCE GROUPS

User notes:

About this appendix: *Appendix B allows the adjustment of performance groups based on occupants or the unique features of a building.*

Table B101.1 can be used for two purposes. If a facility is being evaluated, and there is no clear means of determining a performance group because the facility is not similar to occupancies described in Table 303.1, Table B101.1 can be used as a subjective means of evaluating the associated risks. From that evaluation, a performance group can be developed.

A second use of Table B101.1 would be to evaluate the impact of unusual features in a building or facility for which the performance group is identified in Table 303.1.

Code development reminder: *Code change proposals to this appendix will be considered by the IBC—General Code Development Committee during the 2025 (Group B) Code Development Cycle.*

SECTION B101—RISK FACTOR

Scan for Changes

fda2d49

[BG] B101.1 General. Table B101.1 shall be used as a guide for determining the appropriate performance group allocation for specific structures that have unique characteristics.

[BG] TABLE B101.1—WORKSHEET FOR ASSIGNING SPECIFIC STRUCTURES TO PERFORMANCE GROUPS

RISK FACTORS	RELATIVE LEVEL OF RISK FOR SPECIFIC STRUCTURE
Occupant Load. Maximum number of persons permitted to be in the structure or a portion of the structure.	
Duration. Maximum length of time that the structure is significantly occupied.	
Sleeping. Do people normally sleep in the building?	
Occupant Familiarity. Are occupants expected to be familiar with the building layout and means of egress?	
Occupant Vulnerability. What percentage of occupants, employees or visitors is considered to comprise members of a vulnerable population?	
Dependent Relationships. Is there a significant percentage of occupants or visitors who are expected to have relationships that may delay egress from the building?	
HAZARD FACTORS	
Nature of the Hazard. What is the nature of the hazard, and what are its impacts on the occupants, the structure and the contents?	
Internal or External Hazard. Is the hazard likely to originate internally or externally or both?	
LEVEL OF IMPORTANCE	
Population. Are large numbers of people expected to be present?	
Essential Facilities. Is the structure required for emergency response or post-disaster emergency treatment, utilities, communications or housing?	
Damage Potential. Is significant risk of widespread and/or long-term injuries, deaths or damage possible from the failure of the structure?	
Community Importance. Is the structure or its use largely responsible for economic stability or other important functions of the community?	
SPECIFIC ADJUSTMENTS	
Are the design performance levels adequate and appropriate for the specific structure?	
OVERALL RISK, HAZARD, IMPORTANCE FACTORS & PERFORMANCE GROUP ASSIGNMENT	

APPENDIX

C

INDIVIDUALLY SUBSTANTIATED DESIGN METHOD

User notes:

About this appendix: *When the design analysis and methodology are not based on authoritative documents or design guides, the method of validation in Appendix C may be used in lieu of code provisions. This code requires a peer review in Section 103.3.4 for any methods falling into this category.*

Code development reminder: *Code change proposals to this appendix will be considered by the Administrative Code Development Committee during the 2025 (Group B) Code Development Cycle.*

SECTION C101—GENERAL

Scan for Changes

a08ba36

[A] C101.1 Scope. This appendix is intended to assist in the application of Section 103 where a particular method is not considered a *design guide* or *authoritative document* as defined in Chapter 2.

[A] C101.2 Criteria. Individually substantiated design methods shall comply with one or more of the following:

1. A process to evaluate design options against the performance objectives and functional statements shall be provided.
2. A comparison, signed and sealed by the *registered design professional in responsible charge*, between the prescriptive requirements and this design method shall be provided.
3. *Peer review* shall be provided.
4. This method shall not negatively impact the remainder of the building that complies with the *prescriptive codes*.
5. This method shall address the actual use of the building, including but not limited to the number of people, fuel load, awareness and mobility of the people.
6. The methodology for validation of this method for the project shall be acceptable to the *registered design professional in responsible charge* and the code official.

Where applicable to the proposed design, individually substantiated design methods shall comply with the following:

1. Reports prepared by the evaluation services shall be documented.
2. The data substantiating the building performance as a whole shall accompany the design solution.
3. Where multiple scenarios are applicable, the method shall be substantiated by an approach using not less than two scenarios acceptable to the code official design objectives and code provisions.

APPENDIX

D

QUALIFICATION CHARACTERISTICS FOR DESIGN AND REVIEW OF PERFORMANCE-BASED DESIGNS

User notes:

About this appendix: *Appendix D is provided as a resource to anyone undertaking a performance-based design or review to assess the qualifications of those performing the task. The goal of this appendix is for the tasked design professionals, special experts and competent reviewers to meet technical qualifications in education and experience associated with performance-based design. These qualification characteristics define the level of knowledge or expertise necessary to create or review a performance-based design.*

Code development reminder: *Code change proposals to this appendix will be considered by the Administrative Code Development Committee during the 2025 (Group B) Code Development Cycle.*

SECTION D101—GENERAL

[A] D101.1 Scope. In order for anyone to assess and verify that all of the members of a design team have the knowledge and characteristics needed to execute or review a *performance-based design*, the following lists are provided. This technique is designed specifically for performance-based projects and does not apply to prescriptive-based designs. It is important to understand that utilizing this technique relies heavily on the personal ethics of each individual, and a more formal declaration of education, training and experience may be requested by the code official. These characteristics explain the level of expertise necessary to form a complete design team, but they are not a requirement for every member of the team.

[A] D101.2 Registered design professional in responsible charge characteristics. *Registered design professionals in responsible charge* shall possess the following qualifications:

1. Registered *architect* or *engineer* by the state or jurisdiction.
2. Knowledge of all facets of the project and the underlying principles of the performance-based code and concepts.
3. Ability to perform in the role of point of contact and to coordinate activities between the design team members, owner and code official.
4. Ability to ensure that all elements of submittal to the code official are compatible, coordinated, logical, complete and comprehensive in documentation.

[A] D101.3 Registered design professional characteristics. *Registered design professionals* shall possess the following qualifications:

1. Knowledge of underlying principles of performance-based code and concepts.
2. Education, training and experience in performance-based engineering design.
3. Skill in risk- and hazard-assessment tools as a design method.
4. Ability to utilize performance-based code objectives and to demonstrate compliance through documentation of decision making and solutions.
5. High skill level in engineering disciplines needed in performance-based designs for structural, mechanical and fire-protection systems.

[A] D101.4 Special expert characteristics. *Special experts* are those individuals who possess the following qualifications:

1. Individual has credentials of education and experience in an area of practice that is needed to evaluate risks and safe operations associated with design, operations and special hazards.
2. Licensing or registration where required by a state or jurisdiction for the function to be performed.

[A] D101.5 Competent reviewer's characteristics. The principal reviewer or code official is responsible to acquire competent reviewers with these characteristics and to utilize registered individuals where required by a state or jurisdiction. These characteristics are applicable to the code official's staff and/or contract reviewers. See Sections 102.3.6.2 and 102.3.6.3.

1. Knowledge of underlying principles and concepts of performance-based code provisions.
2. Education in performance-based engineering principles.
3. Competence in risk- and hazard-assessment tools as a design method.
4. Ability to verify *construction documents*, meet analysis and documentation requirements, and to demonstrate that objectives are met.
5. High skill level in engineering disciplines needed in performance-based designs for structural, mechanical and fire protection systems.

APPENDIX E
USE OF COMPUTER MODELS

User notes:

About this appendix: *Appendix E gives guidance regarding qualifications and information that should be provided when undertaking computer modeling. More specifically, the appendix requests that computer program data be submitted as part of the documentation. Also, limitations and applicability of the model must be included as part of the documentation. Finally, the scenarios used to run the particular model must be justified.*

Code development reminder: *Code change proposals to this appendix will be considered by the Administrative Code Development Committee during the 2025 (Group B) Code Development Cycle.*

SECTION E101—GENERAL

[A] E101.1 Scope. This appendix provides guidance on the appropriate use of computer models.

SECTION E102—REQUIREMENTS

[A] E102.1 Use and documentation. The following are issues that shall be addressed where computer models are used in the design of a building or facility.

1. Computer modeling work is required to be conducted under the guidance of the *registered design professional*. Although states or jurisdictions may not require licensing or certification for a computer model operator in areas such as fire, structural, mechanical and energy, knowledge and experience is needed in the application of the program limits and the *performance-based design* objectives for compliance with performance-based code objectives.
2. Computer program data shall be submitted as part of documentation and shall include but not be limited to program name, brief description, type of analysis and application program input and output units and description, and how it is to be used to support design. Statements of exact mathematical model(s) and accompanying submodel(s), if any, uncertainty, assumptions, limitations, scope of applicability and a few reproducible simple benchmark cases shall be included.
3. Background data must be submitted to substantiate why particular scenarios are rejected or accepted.

SECTION E103—RESPONSIBILITY

[A] E103.1 Registered design professional. The computer modeling approach is merely a tool for high-speed calculations that provides mathematics calculations, graphical and related results. It is the *registered design professional's* responsibility to incorporate the above data and background information required as documentation for his or her design document submittal. See Section 102 for more information on documentation.

INDEX

2024 INTERNATIONAL CODE COUNCIL PERFORMANCE CODE® FOR BUILDINGS AND FACILITIES

USER'S GUIDE

USER'S GUIDE
TABLE OF CONTENTS

FOREWORD

Introduction

This User's Guide is provided as background to discuss the rationale and basis for the code provisions. This User's Guide is not considered part of the *International Code Council Performance Code® for Buildings and Facilities* (ICCPC®) but is provided for support with regard to interpretation and background information.

This User's Guide provides an overview of the structure and content of the ICCPC. Additionally, it provides insight about how a performance-based code works, and it explains the particular provisions found within the code. The first edition of the ICCPC, published in 2001, was drafted jointly by two committees composed of representatives from the code enforcement community (building and fire), academia, research firms, design firms and professional organizations. The drafting process consisted of the release of three reports that were open to comments from any interested parties. Following these, another draft was released, officially titled "Final Draft to the *International Code Council Performance Code for Buildings and Facilities*," which was then subjected to the ICC code change process in 2001, resulting in the 2001 edition. There have been eight editions subsequent to the 2001 edition: 2003, 2006, 2009, 2012, 2015, 2018, 2021 and 2024.

The ICCPC is not intended to be any different in scope than the current International Codes® (I-Codes®); however, there are more codes in the I-Code family than when this document was originally developed. It is, in part, a formalization of the alternate materials and methods section of the current I-Codes. Currently, alternate designs occur with minimal guidance regarding the intended outcomes. The current alternate materials and methods approach requires equivalency but does not describe how equivalency should be demonstrated, nor does it provide an administrative process to follow. A performance-based code provides structure to the alternate materials and methods approach and more understanding on the expected outcomes. As part of that structure, the code and User's Guide provide code officials, designers and owners many approaches and resources that can be helpful in managing projects and administering approvals of performance-based designs. It should be noted that some portions of the approvals process found in Chapter 1 of the ICCPC are found in Appendix O of the *International Building Code®* (IBC®), Peformance-Based Application.

This process goes beyond the formalization of the alternate methods and materials section, as it structures the prescriptive code in a more appropriate manner that focuses the code user on the intended outcomes instead of on prescriptive solutions. For example, a subject not explicitly covered by the prescriptive codes is public welfare related to continuity of mission or the overall economic good (welfare) of a community. Examples might include a hospital, school, emergency services center, factory employing a vast majority of the residents of the community or a large mall that represents a significant tax base to the community. For one reason or another, these buildings are necessary for the well-being and viability of a community and thus need to be kept operating or have their "down time" minimized so that they can continue to serve the community even after suffering a major event such as an earthquake, fire or hurricane.

This User's Guide is to be used in conjunction with the ICCPC and not as a substitute for the code. The User's Guide is advisory background information only. The code official alone possesses the authority and responsibility for interpreting the code.

Background

The methodology employed in performance-based codes focuses on outcomes. In other words, a performance code approach would identify and quantify the level of damage that is acceptable during and after a fire, earthquake or other event. Generally, but not in all cases, the current prescriptive code focuses on solutions that achieve a certain outcome. The difficulty is that the outcome is unclear. Therefore, when a design is proposed that is different from the prescriptive code, it is often difficult to determine whether the approach will be equivalent. There may be other more appropriate and innovative solutions available. A performance-based code creates a framework that both clearly defines the intent of the code and provides a process to understand quantitatively what the code is trying to achieve. Without this framework, new construction techniques and innovations would be fairly difficult to accomplish and new methods of construction take longer to implement.

This code also addresses issues that are not specifically related to a natural or technological hazard event. For instance, providing equal access for those with disabilities is not related to an event but instead is a societal expectation of equality. Another example is energy efficiency, which is currently an important expectation of society but cannot be linked to a particular hazard event.

As noted, the prescriptive code is a solution that we have been applying over the years to achieve a certain outcome, and it will continue to be used as the primary viable solution. In fact, most designs under a performance-based code system will be conducted using prescriptive codes. More specifically, a performance code will not replace prescriptive codes. Developing a performance-based code creates a framework in which numerous design solutions are available, including the current prescriptive codes.

The development of performance criteria and acceptance methods is outside the scope of the ICCPC currently and will require representatives of industry, professional societies and research and evaluation services to take a role in the development and application of these criteria and methods. Initially, such criteria and methods were not intended to be found or to be directly referenced in the performance code. The code provides criteria to help determine which methods are acceptable within Section 103, Acceptable Methods. However, there are some specific references now included in the ICCPC within Chapter 5 addressing structural requirements. Chapter 5 references Chapter 16 of the IBC for structural loads and ASCE 7 for target reliabilities.

Structure of the code provisions

This code is divided into three parts as follows:

Part I—Administrative (Chapters 1–4).

Part II—Technical Requirements (Chapters 5–19).

Part III—Appendices (A–E).

Part I—Administrative

Part I of the document contains four chapters in which common approaches were found for both building and fire. Chapter 1 contains administrative provisions such as intent, scope and requirements related to qualifications, documentation, review, maintenance and change of use or occupancy. Also, provisions for approving acceptable methods are provided. Chapter 2 provides definitions specific to this document.

Chapter 3, "Design Performance Levels," sets the framework for determining the appropriate performance desired from a building or facility based on a particular event such as an earthquake or a fire. Specifically, the user of the code can more easily determine the expected performance level of a building during an earthquake. In the prescriptive codes, the required performance is simply prescribed with no method provided to determine or quantify the level of the building's or facility's performance. In other words, all of the different requirements such as heights and areas, sprinklers and structural requirements are attempting to address the hazards to which buildings are subject and the losses that society is able to tolerate. Because these issues are dealt with implicitly, it is difficult to measure the level of safety provided. Therefore, when applying the alternate materials and methods approach for the prescriptive code, it is unclear what is meant by "equivalent," and often the designer must try to make the determination. The problem with the designer determining the intended performance level is that such decisions may not be technical in nature. They are value judgments, which should ultimately be made by policy makers. This chapter can serve as the link between the policy makers and the designers by specifically providing measurable guidance as to desired performance. See the User's Guide for Chapter 3 for a more detailed discussion. It should be noted that the structural provisions within the IBC are somewhat performance-oriented in that buildings are placed into risk categories. See Table 1604.5, Risk Category of Buildings and Other Structures, of the IBC. The structural requirements are then based on those occupancy categories. This table is duplicated into Chapter 3 of this code as Table 303.1.

Chapter 4 deals with the topics of reliability and durability and how these issues interact with the overall performance of a building or facility over its life. This issue has always been relevant to codes and standards but becomes more obvious when a performance code requires a designer to regard buildings as a system. Also, there is often a concern that when performance designs are implemented, necessary redundancies may be removed. As an example, greater dependence may be placed on the use of a single, active fire protection system rather than relying on a combination of passive compartmentation and active fire protection systems. Reliability includes redundancy, maintenance, durability, quality of installation, integrity of the design and, generally, the qualifications of those involved within this process. More discussion is found within the User's Guide for Chapter 4.

Part II—Technical Requirements

Part II of this code provides topic-specific qualitative statements of intent that relate to current prescriptive code requirements.

Generally, the topic-specific qualitative statements are the basic elements missing from the prescriptive codes. The statements follow a particular hierarchy, described as follows:

Objective. The objectives define what is expected in terms of societal goals or what society "demands" from buildings and facilities. Objectives are topic-specific and deal with particular aspects of performance required in a building, such as safeguarding people during escape and rescue.

Functional statement. The functional statement explains, in general terms, the function that a building must provide to meet the objective or what "supply" must be provided to meet the "demand." For example, a building must be constructed to allow people adequate time to reach a place of safety without exposure to untenable conditions.

Performance requirement. Performance requirements are detailed statements that break down the functional statements into measurable terms. This is where the link is made to the acceptable methods.

Societal goals are difficult to determine but need to be reflected within the code, since they are the purpose for having regulations for buildings and facilities. Society expects a certain performance from buildings and facilities and demands that the local codes and their enforcement provide that protection. As noted earlier, such goals need to match the expectations of the policy makers. These goals will vary among communities because of specific needs and concerns such as the preservation of an historic part of a community or a business that employs a majority of the town's work force. Policy makers have relied on model codes to reflect these goals, but model codes today generally focus on the protection of life and property versus looking at a community overall and its unique features. Thus, the desired goals are not always achieved by the simple adoption of model codes. Thus, variations in a community's social objectives are reflected by local amendments. In the performance-based code, objectives, functional statements and performance requirements are generalized by using terms such as "reasonable," "adequate" or "acceptable." In the current prescriptive code, often there is only one value that is deemed "reasonable"; thus, communities must amend the code to reflect their local needs. Justifying amendments is often difficult in a prescriptive code environment since we are looking at single or limited solutions versus understanding outcomes tolerated by society in events such as earthquakes. The performance code is an attempt to create an environment where "reasonable" is qualified by the level of damage that is tolerable to a community based on the type of events expected and the use and importance of the building impacted. It is hoped that this code will create a framework that policy makers can use to reflect what society expects more clearly and in a more consistent way from jurisdiction to jurisdiction.

Part III—Appendices

Part III contains the appendices to the code document. Each of the appendices relates to specific provisions of this code and are discussed within the User's Guide as applicable.

Equivalency versus performance-based codes

The ICCPC is intended to provide a more comprehensive structure than model codes regarding alternative materials, design and construction methods, and testing. Typically, these alternatives were used for specific building or facility applications, such as exiting requirements or innovative techniques for seismic design. Unfortunately, building officials, fire officials and appeals boards find themselves in the position of determining acceptance criteria with only general guidelines.

Performance-based designs have been occurring under the preceding sections for years, despite the lack of guidance to designers or code enforcers. These sections do not address specific subject areas or state specifically what is intended. By contrast, a performance-based code will clearly state the intent for each specific area the code intends to cover, such as means of egress and indoor air quality, essentially expanding on the alternative methods and materials section. Thus, decision making in terms of methodologies beyond the prescriptive code may be clearer. The intent statements are satisfied through the use of acceptable methods, which include the prescriptive code. The methodologies put forth in this code provide a structure in which the designer and the code official are provided with the flexibility to determine a specific design performance level desired by the owner based on an evaluation of the level of risk and reliability of the solutions used, while still achieving the intent of the code.

It is anticipated that prescriptive code provisions of the IBC, *International Residential Code*® (IRC®), *International Mechanical Code*® (IMC®), *International Plumbing Code*® (IPC®), IFC and related codes will likely be used for most projects, and performance-based designs or a combination of prescriptive and performance-based designs will only be used for unique circumstances requiring design innovation. A benefit of the ICCPC, though most designs will be accomplished with prescriptive codes, is an improved understanding of the intent of the prescriptive requirements. Ultimately, the improved understanding can benefit the quality of the prescriptive documents by providing additional solutions generated using a performance code structure. Such a document can also serve as a sounding board for the inclusion of requirements into the I-Codes during the code change process.

"Architecture" of the codes

There are often questions related to how the document will work with prescriptive codes in the adoption process. It is intended that this performance code is to stand alone from the prescriptive codes but utilize the prescriptive codes as an acceptable method. Essentially, if a jurisdiction were to adopt the ICCPC, it would still be using the prescriptive codes, such as the building, residential, fire, mechanical and plumbing codes, as acceptable methods. Also, it will always be possible that a jurisdiction could adopt the prescriptive codes without adopting the ICCPC. The ICCPC is intended as a framework document that creates a method more closely reflecting society's expectations of building and facility performance. This document has the necessary components to reflect society's expectations through better communication of intent and through placing buildings, structures and facilities into more conservative performance groups.

Past experience with prescriptive codes, specifically with regard to seismic design, has demonstrated that public expectation of regulations is often higher than what the regulations and technical communities provide. The Northridge, California, earthquake on January 17, 1994, was an example of this disconnect. The engineers were satisfied that most buildings met the objective of life safety, whereas the public expected buildings to be much more usable after the event. Additionally, it is recognized that it will be some time before an overall performance-based system is adopted by most jurisdictions.

Existing buildings and facilities

A performance-based regulatory system can be used as a tool in understanding at what level an existing building may perform in an event such as a fire or earthquake. There are some prescriptive tools available in current codes, such as the work area method and scoring method found in the *International Existing Building Code*® (IEBC®). These methods have become fairly developed and provide a fair amount of flexibility but still do not indicate the actual performance of the building or portion thereof. Additionally, fire codes in the United States have limited requirements related specifically to existing buildings and facilities, but in cases where it can be shown that a distinct hazard exists, the requirements for new buildings and facilities would apply. The burden of showing that a distinct hazard exists rests with the jurisdiction. A performance code can be used to better understand the level an existing building or facility may perform in events such as a fire or an earthquake, for example. This could apply to a single building, a certain occupancy classification or to an entire community through an adopted ordinance. The ICCPC provides a tool to the jurisdiction to help assess the level of performance of existing buildings or perhaps a building type.

Additionally, a performance code provides options to designers and owners when they address a hazard in an existing building. For example, instead of simply stating that the travel distance must be decreased, the objectives of the code are discussed, which allows designers to determine other ways to achieve the objectives. A systems approach in many cases gives everyone involved a better, more realistic understanding of the actual hazards and the most effective means of addressing such hazards.

Summary

The approach provided in the performance-based document is in part a global expansion and a major improvement on the equivalency approach. Also, the comprehensive structure that includes strengthened administrative provisions, design-performance level decision-making tools and topic-specific intent statements focuses codes on desired outcomes instead of on a single solution.

The prescriptive code is a very important element of a successful performance-based code system and will continue to play an important role in the future. Specifically, a performance code structure will create an environment that encourages innovative approaches, which may become solutions within the prescriptive code.

Currently, the design performance level concept provided in Chapter 3 of the code document specifically focuses on events such as fires, earthquakes and winds, and does not address everyday use issues such as interior environment, prevention of falls and various other topics that could be dealt with on varying design-performance levels. If Chapter 3 is not referenced within a specific chapter in Part II, then there is assumed to be only one design performance level.

Some additional issues that need further study are durability and reliability and how these affect building performance.

CHAPTER 1

GENERAL ADMINISTRATIVE PROVISIONS

SECTION 101—INTENT AND PURPOSE

Scope

The scope statements encompass all portions of the code and are similar to the corresponding administrative provisions of the prescriptive codes. They provide an overall understanding of the limits and applications of the document. For example, the scope statement reads in part: "This code provides requirements for buildings and structures and includes provisions for structural strength, stability, sanitation, means of access and egress, light and ventilation, safety to life and protection of property from fire and, in general, to secure life and property from other hazards affecting the built environment." The purpose of the performance code is to promote innovative, flexible and responsive solutions that optimize the expenditure and consumption of resources while preserving social and economic value. This approach is unique to the structure of a performance-based code. This code also provides the opportunity to better understand how all objectives are being met. Essentially, it is a more holistic approach than the prescriptive codes. In some ways it is the next generation of the alternative methods and materials allowance in the prescriptive codes.

The intent statements have been expanded beyond the traditional intent of the prescriptive ICC family of codes to address varying levels of performance. In addition to these intent statements, specific goals in the form of objectives are found within each individual chapter to further delineate for all stakeholders the more detailed intended performance for the various goals mentioned in the intent statement such as egress and energy efficiency.

It is important to understand that there are requirements in the form of objectives, functional statements and performance requirements applicable to buildings or structures that must be addressed. This code could apply to buildings, but also to outdoor hazardous materials storage or processes. The lifecycle of these issues, such as what would be expected in a fire code or property maintenance code, are integrated throughout this code with reliability, durability and other long-term, operational and process-related requirements. In addition, this code can be applicable to existing buildings and facilities.

SECTION 102—ADMINISTRATIVE PROVISIONS

The administrative section discusses how the performance code works in terms of the practical application of the code including stakeholder qualifications and responsibilities, document submittals, and review and construction verification techniques to demonstrate that the performance code objectives have been satisfied. Additionally, this section emphasizes the importance of the long-term maintenance needs of a building that has used a performance-based design approach. Any changes must be appropriately managed, whether such changes are large or small.

The overall regulatory process regarding the administration of the code enforcement program (as outlined in Chapter 1 of the prescriptive ICC family of codes), appointment of the code official and staff, basic permit requirements and exceptions, fees, inspection types and requirements, and appeals is not included in the performance code administrative provisions for the following reasons:

- Requirements of the ICC family of codes are to be utilized as specifically referenced in the performance code provisions.
- A major emphasis of this code is the unique administrative approach necessary for successful performance-based designs.
- Performance-based design methods are, in most cases, only expected to be applied to portions of a building, structure or facility, or to process certain types of contents within a building; therefore, supplemental requirements to the prescriptive codes will be necessary.

The section also addresses various responsibilities shared by the partners in a performance-based design, ranging from the building or facility owner to the registered design professionals, to the code official. The term "code official" is used multilaterally to incorporate building, fire, mechanical, plumbing and any other officials as may be designated by the adopting jurisdiction.

The model codes have traditionally incorporated alternate materials, alternate designs and alternate methods of construction. This code provides a more holistic framework to utilize new materials and methods when design equivalence to the prescriptive code or achievement of objectives is demonstrated and accepted by the code official. This code provides a more disciplined, comprehensive approach with the intent of the code and the objective of each chapter clearly identified and administrative requirements clearly stated. The expectation is that design professionals should use this code as a tool to improve the quality of written submittals. Further, the comprehensive structure of a performance code diverts the focus away from a single solution and instead emphasizes code intent and objectives, allowing for design solutions appropriate to a particular situation as opposed to designing to the minimum or only focusing on very specific code requirements. This code provides a structured and comprehensive method to facilitate innovative designs. This approach requires more comprehensive written submittals by design professionals, requiring documentation of project objectives, design compliance with code objectives, design basis using authoritative documents, and analysis.

Other resources include the following:

- *Code Official's Guide to Performance-Based Design Review* (SFPE and ICC, 2004)

- *Performance-Based Building Design Concepts, A Companion Document to the International Code Council Performance Code for Buildings and Facilities* (ICC 2004)
- *SFPE Engineering Guide to Performance-Based Fire Protection Analysis and Design of Buildings* (NFPA and SFPE, 2007)

102.1 Objective

The goal of the administrative chapter is to achieve and maintain safety through clarification of the responsibilities of the owner, registered design professional and code official; to provide requirements for preparation and submittal of performance-based design documents; and to provide for methods of verification and documentation as a systems approach to comply with the performance code objectives over the life of the design. This includes accountability of all stakeholders with respect to design, implementation of the design, maintenance and management of future changes to the use, facility or building.

It is important to note that regardless of how a building or facility is designed, ultimately the design must translate into practical construction documents that can be used for the construction process. This chapter focuses on the process and overall documentation, but it is always the intent that clear direction is provided for the construction process. Generally, constructions documents are discussed in the functional statements.

102.2 Functional statements

The functional statements set forth the steps necessary for a successful performance-based design process. These steps range from the required technical competence of the design professionals, the documentation essential for a competent review and approval, the implementation of the approved performance-based design and verification of compliance, the long-term maintenance of the original design and the careful management of any changes.

Reference is made as well to the administrative provisions of the International Code Council's family of codes in regard to the dynamics of the agency's plan review, permit issuance, inspection procedures and enforcement policies. Any additional administrative procedures should be considered for inclusion in the adopting ordinance or regulations of the jurisdiction.

102.3 Performance requirements

This section is divided into code subsections by particular subject matter. The purpose is to focus on responsibility and methods of accountability regarding major activities. The methodology of managing performance-based design submittals, review, construction, inspection and testing relates to other code's practices but is expanded within this code based on several successful building department programs currently in place. Good documentation and the use of a quality assurance program for large or unique construction projects are highly recommended tools for managing performance-based design projects.

Emphasis is placed on building and facility maintenance to retain compliance with the performance-based design. Future remodeling, additions and changes of use are perpetual factors for existing buildings and require solid documentation to aid the design professional's evaluation of a building and any associated use changes in the future. The administrative objectives are considered requirements related to responsibility, process and methods of verification. It is critical that all the elements work together to meet code objectives. A higher level of responsibility, accountability and documentation is required for performance-based designs to achieve safe buildings.

102.3.1 Building owner's responsibility

Prior to preliminary design development, it is necessary that the owner, through the principal design professional, secure a qualified design team of experts with experience in performance-based design techniques. A building or facility owner's costs may be substantially increased as a result of using a performance-based design, but the potential for creative design and a unique and functional building or facility may be obtained to meet the owner's objectives.

It is strongly recommended that all pertinent issues be discussed candidly among all stakeholders during the preliminary design phase to achieve "buy-in" by all parties and that a review of the higher-level objectives found in this code be undertaken. Stakeholders include the owner, operators, principal and other design professionals, code officials, contractors, representatives of the lenders, insurance and other firms who can have an impact on the design and use of the building. See the User's Guide for Section 102.3.1.1 for a discussion of the considerations for the selection of this code.

A building or facility owner has a major responsibility when he or she chooses a performance-based design approach. This responsibility must be explored in depth from the conceptual and preliminary design forward, specifically examining how the owner or operator will operate and maintain the building in accordance with the approved design. The owner or a suitable representative should be actively involved in the design process, including meeting with the code official, to ascertain immediate and long-term responsibilities. If owners choose not to accept this responsibility, they will probably opt to direct their design professionals to utilize current prescriptive codes and standards.

102.3.1.1 Registered design professional

Traditionally, for prescriptive code projects, a project owner hires the design professionals, establishes the project vision, outlines the functional project criteria and provides instructions to the design team. On projects where more than one design professional is hired individually without having responsibility to one single design professional (architect or engineer) in charge, code officials have likely encountered cases where design documents were not coordinated and other cases where the multiple design professionals worked toward different objectives. This process can result in design documents that require substantial revisions before plans can be approved to comply with the minimum standard of prescriptive codes. This leads to systems that are not compatible and construction, operational and maintenance problems.

In a performance-based code, the practice above is not acceptable and steps must be taken on the front end of a project to ensure that all design work is coordinated and meets the code and design objectives. This can be overcome by the owner acquiring design services through one party, with one design professional having contractual responsibility and authority to acquire services from other design professionals, to provide direction and coordination, and to verify that all design professionals are working as a team in the design process.

The owner's action to hire a principal design professional as a single party who will direct and coordinate all design activities is the most critical prerequisite when selecting to apply the performance code to a specific project. The first step in this critical process is to get the owner's and design team's joint buy-in on several objectives of this code (for example, establishing bounding conditions, developing an operations and maintenance manual, acknowledging the building owner's responsibility of code provisions and convincing the code official that the application of the performance code will result in compliance with the objectives of the code).

After the principal design professional, other design professionals and the owner have thoroughly discussed the owner's vision, objectives and design concepts that require the application of this code versus a prescriptive code, it is necessary that the principal design professional assess the design team to determine if additional skilled design professionals are needed to begin evaluating the performance code and its application. After the design team has scoped the project design requirements and assessed the programming and schematic phase of the proposed project, a thorough discussion is needed among the owner, the principal design professional and the design professional of each design discipline to obtain buy-in and understanding of their respective responsibilities. It is recommended that the team review the proposed project at this time with the code official to verify that the developed proposal will work within the code objectives and to assess its application to the project. Also, the code official's feedback and peer review feedback, where applicable, are necessary relative to the code official's acceptance of the conceptual proposal to date and the use of this code. The code official should be involved with a project's design team from the conceptual stage through completion of construction to assure input and buy-in of the design approach among both the members of the design team and the code official. The intent of the concept report and design reports listed in Section 102.3.4.2 of the code is to ensure formally that communication and buy-in are maintained throughout the design phase of the project. Further, it is recommended that the owner also include the contractors, lenders and operating staff to obtain buy-in of a performance-based design project.

102.3.1.2 Registered design professional in responsible charge

The building, facility or project owner retains the services of a registered design professional; establishes the project visions, goals and functional needs; and places primary responsibility for the overall design with a registered design professional in responsible charge. One of the key elements highlighted within the performance requirements is the need for overall coordination of documents by the registered design professional in responsible charge as designated by the owner. The success of a performance-based design lies in the development of a performance solution to reach the desired objectives. Since there is often more than one design professional as a member of the design team, the owner should empower a single registered design professional in responsible charge with the authority to coordinate the design professionals in order to achieve a coordinated and complete performance-based design.

102.3.1.3 Peer review

To assist in the quality assurance process, the code official can require an independent design professional or recognized special expert to perform a peer review of one or more components of the performance-based design. If the code official does require such a review, the owner is obligated to furnish it at his or her expense. The peer reviewer plays an important prescreen role for the jurisdiction and serves as a critical review on assumptions, design approaches, hazard scenarios and other technical aspects of the design prior to local agency submittal. Peer review is most effective at the beginning of the development of designs, preferably long before proposed construction documents are submitted to the code official.

102.3.1.4 Costs

As stated in Sections 102.3.1.1 and 102.3.1.3 of the code, the owner is responsible to acquire and fund all the design and support services including peer reviews, special experts and field observation by the principal and other design professionals. Section 102.3.1.4 covers any other additional responsibilities, such as contract review as required by the code official.

102.3.1.5 Document retention

Once a performance-based design has been approved and implemented, the owner is obligated to retain all documentation required on the premises and available for review by the code official. This may be in electronic form as approved. Post-implementation inspections might raise any number of questions concerning the assumptions, bounding conditions or other aspects of the original design. Part of the design documentation, more specifically the operations and maintenance manual, becomes the inspection handbook for that building or facility or part thereof. Changes to the original design, when proposed or noted on verification tours, must be evaluated in conjunction with this chapter and the original design documents to determine the proper course of action. Electronic storage and management tools will be beneficial for such document retention and updating in the future.

102.3.1.6 Maintenance

Maintenance of the performance-based design is perhaps the most challenging aspect of performance codes. Performance-based solutions may, at times, deviate substantially from traditional building and fire codes. Although the fire department or code agencies may inspect such buildings or facilities for compliance, this section provides a better understanding for building and facility owners as to limitations and features in terms of the risk and hazards that may result based on their action or inaction. For example, an arena event with a large number of small children may require more security personnel or crowd managers than other events given the associated occupancy characteristics or the technical or behavioral hazards. Performance-based solutions often provide an understanding of the true hazards of certain occupancies when the specific use is more closely evaluated.

102.3.1.7 Changes

A second aspect of maintaining a design revolves around changes that normally occur within buildings and facilities. Section 102.3.1.7 clearly prescribes that the owner shall ensure that all changes, including processes and systems, do not increase the hazard level beyond the level established in the original design. A second but equally important concept is a requirement that all changes impacting the performance-based design be documented and made available for review by the code official.

102.3.1.8 Special expert

This code allows for the use of special experts (characterized in Appendix D) when the scope of work is limited to that which does not result in the practice of architecture or engineering. There is a significant amount of potential performance-based design work, such as in the fields of hazardous materials, contents and process safety management, and fire protection design, that is exempt in many jurisdictions from practice laws. These special experts may well be the most effective resource in these specialized areas. To impart some minimum discipline in the use of these experts, it is the stated intent of the code that such individuals or firms meet the qualifications listed in Appendix D.

102.3.1.9 Occupant requirements

Where the life of a performance-based design relies to any degree on human decisions or actions, the owner of a building or facility has a continuing obligation to provide all occupants and employees who may have a role in the performance-based design, however limited, with the requisite training to develop the skills to undertake those roles and empowerment to apply such training. For example, a performance-based design that relies on a staff-coordinated assisted evacuation requires development of a plan and attendant drilling. The owner is obligated to be sure that such training is kept current and that all new, temporary or replacement employees are promptly provided with the information and training needed to effect the response. The notion of empowerment speaks to situations wherein the staff that is expected to perform an action in an emergency is unable to do so by rules or by being placed in a position, physically or otherwise, effectively disabling their response. If such a situation has a reasonable likelihood of developing, the design team should de-emphasize the human response factor in favor of other strategies.

102.3.2 Registered design professional qualifications

The qualifications of registered design professionals—their knowledge of design, analysis, research, computation, documentation and professional standards in their respective areas of expertise—are prerequisites to a successful performance-based design approach. Qualifications are regulated by professional registration laws within the state or jurisdiction in which the project is to be constructed. Appendix D of the code lists specific qualifications for design professionals involved in a performance-based design. These characteristics detail the level of expertise necessary in a complete design team but are not requirements of each individual team member. In fact, many designs related to Chapter 19 of this code may well be outside the scope of practice laws and may be performed by a special expert. The special expert, too, must possess the qualifications listed in Appendix D.

In addition to the registration requirements that may exist within the jurisdiction, the code official may require a formal declaration of education, training and experience of all design team members to ensure that sufficient expertise has been retained in all technical subject areas of the performance-based design. The performance-based solution is a systems approach, and all components must be addressed with a proper level of sophistication that can span many design disciplines.

Registered design professionals and other members of the design team (such as architect/engineer staff; interior, theater and kitchen designers; modelers; and computer support staff) have a responsibility to gain technical skills through professional and/or university programs before embarking on a design project applying performance-based design techniques. The intent is that the design team must consist of sufficiently qualified individuals to carry out the responsibilities necessary to design the proposed project.

The minimum qualifications and skills of the design professional will vary significantly based on the type of project and the degree of analysis that the project requires. It is safe to say that, as a general rule, a performance-based design will require more analysis than the same design undertaken using a prescriptive code. The main difference is that the decision-making process involves multiple objectives to demonstrate compliance with performance provisions and acceptable methods. It follows, then, that increased project documentation of these many decisions and corresponding solutions will be necessary. Therefore, the qualification and skill levels for design professionals may vary as a function of the size and complexity of the project. It is incumbent upon the design professional to become competent in any areas of practice where new skills are required before undertaking a performance-based design approach.

The increased qualifications and required skills are particularly important for fire protection engineers (sometimes termed "fire engineers"), who will most likely need fire-modeling skills to analyze a sufficient range of scenarios. Design-team leadership skills also need increased attention and enhancement, since the techniques and processes discussed in the acceptable methods of each applicable objective must be met and documented. This will require extensive coordination skills in the leader to enable team members to produce a system design with compatible components. Further, most projects will only be partly based on a performance-design approach; many of the design features will remain prescriptive and must interface with the performance-design components. Ultimately, even a performance-based design approach should result in a prescriptive solution for those involved with construction. Where a design calls for an automatic fire alarm system, an alarm system or components must be specified for installation, including clear direction on layout and spacing.

Large, complex projects using performance approaches in a variety of aspects of the building would most likely require highly qualified and skilled engineers in the structural, electrical, fire and mechanical disciplines.

102.3.3 Registered design professionals' and special experts' responsibilities

Performance-based provisions allow design professionals more freedom to develop innovative solutions and design techniques based on a desired performance as opposed to the traditional prescriptive code provisions that provide nominal bounds on the design. This freedom also establishes a concurrent responsibility for the professionals to utilize the design analysis process that is most appropriate to meet the code's objectives and as required for documentation by the authoritative documents or design guides. The design professional must possess in-depth knowledge or obtain skilled team members to deal with issues such as risks, load factors and occupant life-safety impacts. Each of these issues is addressed in the performance objective statements.

In cases where more than one design professional or discipline is involved in a performance-based design, the owner must designate the registered design professional in responsible charge to ensure the performance design is comprehensive, coordinated and complete prior to submission for review. The registered design professionals of each discipline must work as a unit to provide a systems approach, and a single registered design professional in responsible charge is required to ensure that the coordination of every aspect of the design package is achieved. This section intends that all documents, reports, drawings, calculations and any other relevant materials be submitted through the principal design professional. The goal is to eliminate or at least minimize cases where independent submittals may lack coordination with the overall design and unduly delay or complicate the review process.

The registered design professional in responsible charge is responsible for acquiring the services of design professionals whose qualification characteristics meet a particular level such as those provided in Appendix D. Additionally, design professionals who are in responsible charge of each design discipline shall comply with Appendix D. Performance-based design is intended to be a systems approach, and the registered design professional in responsible charge and the registered design professionals of record for each discipline are accountable and must verify that all components of the design work together to meet the code and design objectives and that verification methods are prescribed in the design to show that the constructed building or structure meets these objectives.

The role of the registered design professional in responsible charge includes the following:

- Coordinate the design professionals of each discipline to ensure that the design methodologies and assumptions are compatible for a systems-approach and that the performance code provisions and any applicable prescriptive code provisions are fully met.
- During the design phase, function as a point of contact for all participants, which includes the design professionals of each discipline and allied consultants, owners, contractors, peer reviewers and code officials, as applicable.
- Ensure that design documents are coordinated and comprehensively complete with appropriate delineation between plans and related documents, and that the submittal contains the necessary support documentation to establish that the design complies with all applicable code provisions.
- Function as a point of contact with the code official to ensure the complete design documents and applications are filed with the government entity for review, approval and permitting. Ensure timely response to questions, revisions and requests for additional information on any element of the submittal.
- Function as a point of contact for the design team following permit issuance or design approval and respond to changes, clarifications or additional information that may be required during the implementation or verification processes.

All design professionals involved in a performance-based design, including the registered design professional in responsible charge, must apply the appropriate performance requirements and select suitable and compatible acceptable methods when using this code. The proper application entails adequate design analysis and support documentation to validate the design approach and to verify that design objectives have been met. This includes research, assumptions, computations and supporting documentation in the form of authoritative documents or design guides that, among other things, determine testing and verification methods. Two critical aspects of the submittal are that the registered design professional references specific documents or design guides, and that the registered design professional clearly demonstrates how the documents and design guides apply to the particular design solutions. Assumptions should not automatically be made that nationally recognized standards demonstrate compliance with performance code provisions. Some standards may apply, but others may not. Additionally, test applications traditionally used by a standard may have to link more appropriately with more realistic performance and the performance code objectives.

Construction management may be critical for an owner in order to ensure code compliance, but such services are generally the option of the owner. Given the complexities and uniqueness of a performance-based design, this code explicitly requires that the registered design professionals review the completed construction elements and systems to verify compliance with bounding conditions and the approved design range of acceptability. The code official is further empowered to require the registered design professional in responsible charge to file a verification report attesting that the design bounding conditions and critical elements designated as part of the approved design have been implemented properly as a condition of issuance of any required certificates.

The design professionals have an ancillary responsibility to the owner in addressing the overall cost effectiveness of the design throughout the life of the building or facility. In some cases, money saved in the initial design and implementation costs results in future costs or a building or facility that is limited in function due to the specificity of the design. For example, a warehouse or process area designed for the storage, handling or use of a particular commodity may or may not be suited to additional quantities or different classes of commodities, unduly limiting the owner's options. This can be the case in the application of prescriptive codes as well. Similarly, an assembly space that is severely limited regarding interior finishes or fuel loads may not adequately accommodate seasonal decorations or events. A realistic analysis of long-term use and maintenance needs must occur to produce an effective performance-based design that will accommodate the owner or tenant's future needs.

102.3.4 Design documentation

Design professionals are responsible to prepare and submit a complete and integrated set of construction documents to the code official, including the following types of information:

- Plans and specifications.
- Calculations to demonstrate that the design analysis meets the standard requirements for professional practices and acceptable methods.
- Computer modeling and analysis with program name and description, program objectives, input and output units, characteristics and related information for plan review verification of compliance with the performance provisions, design objectives and professional standard of practice.
- Assumptions, limitations and factors of safety used in the analysis and design.
- Identity of applicable design components that comply with performance code provisions and the design data to demonstrate compliance, including interface with the prescriptive code provisions on a system basis.
- Identity of design performance levels and objectives determined in Chapter 3.
- Identity of performance criteria utilized.
- Description of scenarios with applicable data used to demonstrate various design approaches and conditions that meet performance provisions.
- Description of the methods used to demonstrate that a standard of care was taken when using a building systems-design approach.
- Scope of inspection and testing requirements to demonstrate compliance with design and code provisions.
- Scope of quality assurance techniques proposed to demonstrate that construction systems comply with the construction documents.
- Commission requirements for final inspection, testing and functional operations to demonstrate compliance with approved design.
- Identity of maintenance requirements and their frequency that an owner must undertake in the future use of the building (for example, inspections, testing, service or maintenance activities).

Proper documentation is essential in performance-based design to clearly identify the objectives of the analysis, the approach or methods taken, the use of automated design and other tools used in the final design, and the substantiation that the design meets the project objectives.

Documentation to support the design should include the following information when applicable and must be maintained for the life of the building:

- Geotechnical, site hazard and other applicable analytical and test reports that provide substantiation or are used in the design.
- Calculations to demonstrate the methods and assumptions of analysis based on recognized analytical methods and mathematical models.
- Technical references to publications that have been accepted by an appropriate professional peer review process.
- Computer analyses supported by documentation that include the program name, a brief description and type of analysis and application, the program output with units and descriptions and how the program is used to support the design. Statements of uncertainty, assumptions and limitations must be included. Reference Appendix E for requirements in using computer models.
- Statements describing methods, techniques and means of verification to demonstrate compliance with performance provisions and how applicable elements are integrated in the design as a system.

- Statement of requirements for testing, inspection and maintenance to identify the basis and criteria for acceptance of the building and applicable components.
- Statement of responsibilities of the building owner after construction to ensure compliance with design parameters.

If utilizing emergency responders as part of a design, the design professional needs to take into consideration the reliability of the organization providing the emergency response service as well as the level of response that can be provided over the life of the building or facility. It is assumed that emergency responders under the direct control of the owner, such as an on-site fire brigade at a manufacturing plant that is staffed by properly trained employees of the owner, can be relied upon in the design process provided the level of response is documented and the key characteristics of the emergency responders are designated as bounding conditions. This approach would require a reassessment of the design should the level of response be reduced by such factors as reduced staffing, less training, longer response times, inadequate equipment, etc., as required by Section 102.3.11.3. However, where emergency response is a service provided by a government entity or public/private organization that is not under the direct control of the owner, a jurisdiction may be reluctant to allow the design to rely on a specified level of response because of the legal ramifications of having to ensure that such level of response will be maintained over the life of the building or facility. The jurisdiction should have the final say on what level of emergency response, if any, the design assumes will be available so that the design will not become deficient should the level of emergency response be reduced by the jurisdiction because of a lack of available funds or for other reasons such as changes in response times, water supply or other conditions beyond the immediate control of the jurisdiction or the organization providing the emergency responders.

The plan review process relies heavily on adequate documentation to demonstrate complete submittal and thorough review of the design to verify code compliance. Performance-based design review will require review of additional documents not required in prescriptive design submittals.

The design documents must also identify when and where special inspection and testing are required to verify compliance. Special requirements shall be specified by the registered design professional in the design documents as requirements for off-site fabrication and on-site construction. All required inspections and tests require verification before the total project can be approved and occupied. The performance codes require verification techniques, which are elements not required for prescriptive code designs due to the overlapping requirement of prescriptive codes. Contractors and inspectors must follow the design documents and not accept deviations to approved plans based on past industry practice.

Documentation begins during the design process as a methodology of tracking and verifying that analysis and application of scenarios of good design practice have been conducted. Documentation should demonstrate compliance with performance code objectives, review and acceptance by code officials, construction inspection and testing, and issuance of certificate of occupancy to demonstrate that occupancy conditions and related community requirements are met. This documentation is necessary during the life of the building to verify that the use and maintenance comply with occupancy conditions. Additionally, the documentation is critical during future remodeling, renovations, additions and changes of use.

Section 102.3.4.2 allows the code official to require (1) a concept report; (2) a design report; and (3) an operations and maintenance manual. These reports compile formal documents for the preliminary design (programming and schematic) phase, design analysis and development of plans and specifications for construction, and future use and maintenance of a project. The reports are intended to improve project communications and conceptual understanding for projects by providing data for each report as follows:

1. **Concept report.** The concept report must include general project information, project scope, building description, occupant characteristics and risks, project goals and objectives, proposed event scenarios, methods of evaluation and proposed acceptance, documentation and qualification of design team, and owner's representatives and contractors. The concept report is to inform the code official of the design concepts and to achieve consensus that the preliminary design approach is acceptable. The conceptual site and building plan should provide sufficient information and setback data so that hazards and separations can be evaluated similar to a code analysis using the prescriptive building code for occupancies, height, area and setbacks.
2. **Design report.** The design report documents project scope, goals and objectives, steps taken for analytical analyses with identification of performance criteria, parameter input assumptions and limitations, magnitude of design loads, computer model data and scenarios for evaluating acceptance ranges, final design, evaluation, criteria design assumptions with bounding conditions, critical design features and final design bounding conditions, commissioning testing requirements with acceptance criteria and supporting documents and references. The registered design professional in responsible charge should provide copies of supporting documents and references necessary for a professional review to be completed. Design guidelines, handbooks and other applicable authoritative documents prepared by professional associations may expand or further clarify applications of the items listed in the User's Guide, Section 102.3.4 and recommend the use of certain performance-based design processes.
3. **Operations and maintenance manual.** The operations and maintenance manual is a formal document for the facility owner and operator that incorporates design, builder and manufacturer requirements in the form of actions that need to be performed on a regular basis to ensure that the components of the performance-based design are in place and operating properly. The manual must identify the restrictions or limitations placed on the use and operation of the facility so that the facility stays within the bounding conditions of the approved performance design. Such restrictions may, for example, include fuel load limits in an assembly occupancy. The manual should contain the following:
    - Limitations on facility operations because of design bounding conditions.

- Identification and description of critical systems.
- Description of required system interactions.
- Periodic maintenance and testing requirements.
- Emergency and typical operational responsibility.
- Staff training.
- Manufacturer's requirements for operation and maintenance of equipment.
- Materials and systems affected.
- Power and utility support requirements.
- Emergency and backup plan for critical component failure.
- A documentation plan for supervision of operations, maintenance and testing of required items in the manual.

The practice of requiring reports at various phases of a project to demonstrate compliance with the prescriptive code has worked very well for several jurisdictions that experience large, complex projects and submittals of alternate materials, designs and methods that include performance-based designs. These reports have demonstrated equivalency or compliance with prescriptive codes and include elements such as fire modeling, fire protection, exiting and other life-safety features, special inspections and testing requirements.

Performance-based design projects are more successful when project criteria and reviews are included from the conceptual stage through design review and construction with an agreed upon, documented quality assurance program. A quality-assurance team approach is recommended during the design and construction process to improve communication and thereby improve the working relationships to obtain a building that meets performance code requirements. The team should include registered design professionals, representatives of the code official for plan review and inspection, a special inspection and testing (third-party) agency and the contractor(s).

Design documents are the basis of design approval and the governing requirement for construction and verification by inspectors and testing agencies for compliance. Deviations from the approved construction documents require the contractor to consult with the architect or engineer to evaluate and approve revisions against the project objectives before obtaining approval from the code official. The process also helps in securing the owner's concurrence for construction to comply with construction documents. Consequently, there is less leeway given to the contractor for not complying with the plans, since the contractor and owner should have a front-end commitment with all the parties.

Standard plan design document submittals are accepted by code officials in many jurisdictions for multiple types of uses within a single jurisdiction. The design professional who is requested to prepare this type of design should contact the code official before initiating a design to determine if the local jurisdiction's procedures authorize standard plan or multiple-use design documents. A design professional proposing a performance-based design using this code for a standard plan should contact the code official to verify that both approaches are accepted and to learn how the jurisdiction's procedures apply to the proposed project.

102.3.5 Design submittal

The registered design professional in responsible charge for projects with multiple design professionals must coordinate all design documentation for compatibility and completeness and ensure that the documentation required in Sections 102.3.2 through 102.3.4 is included in the submittal to the code official. Projects with one design professional must meet the same standards prior to submittal. Design documents must clearly indicate the areas where performance and prescriptive codes apply, so that it is clear to the reviewer which code provisions apply.

Documents should be submitted in accordance with the procedures of the jurisdiction and in sufficient detail to obtain permits for the project. Reports and preliminary documents required by the code official, such as concept reports, design reports and operations and maintenance manuals, may be required earlier than full design document submittal to obtain permits so that conceptual or phase reviews can be made.

102.3.6 Review and approval

Before approving and issuing a permit, the code official or other designated individuals are responsible to review or ensure that construction documents meet the requirements of the performance code as well as the prescriptive code provisions that are being utilized as part of the design. Although performance-based designs will probably be submitted a small percentage of the time, they most likely will complicate the plan review process. Additionally, that small percentage of projects represents those buildings where a significant number of the design systems are performance-based. There will be many designs that are primarily based on the prescriptive codes but incorporate minor performance-based elements.

Performance-based design goes far beyond the traditional design perception that document submittals are automatically acceptable and require little or no review when signed and sealed by a registered design professional. Registration and a license to practice engineering do not necessarily constitute acceptable qualifications to undertake a performance-based design.

The code official will have a higher level of success in the design document review process when the registered design professional has complied with the documentation requirements of Section 102.3.4.

The code official can still undertake review in a traditional manner for a majority of the projects. Section 102 is focused primarily on performance-based design-related projects. It is very important that the code official evaluate the skills of the department before undertaking such a review. Understanding review limitations and addressing these limitations ahead of time will help to ensure a successful and thorough review. This understanding of limitations includes evaluating the educational qualifications and experience needed to review performance-based designs and analyses as stated in Appendix D. The code official may consider one or more of the following options to develop a plan for ensuring qualified plan review:

1. Determine if the staff has sufficient education, engineering or architectural registrations, and levels of continuing education needed to perform the tasks.
2. Initiate a program to upgrade plan reviewers' educational and professional qualifications.
3. Acquire contract review services via a consultant plan reviewer with applicable qualifications.
4. Utilize a peer review process as stated in Section 102.3.6.3 that will provide individuals or a team meeting an appropriate level of qualifications, standards of ethics and accountability.

When the code official accepts a consultant as a contract reviewer in lieu of their own staff's review of construction documents, documentation must be provided for the code official's acceptance indicating that verification of the performance and prescriptive code provisions have been met. This document should become part of the official records for the project and the basis of approving the design documents and issuing a permit(s).

After the design documents and supporting documents and reports are verified as meeting this code and other applicable codes, permits should be issued in accordance with code official procedures.

Costs for review of design documents and for inspections should be the burden of the owner. Since costs for services are expected to exceed the fees collected for typical plan review and inspection services, the code official should amend the local code to include cost recovery for services rendered on all projects using a performance-based design approach. This may also be an issue with alternative methods and material applications related to the prescriptive codes.

102.3.6.3 Contract and peer review

In the design and construction of buildings and structures, the traditional process of analysis and design by design professionals, plan review and inspection by code officials, and special inspection by quality assurance personnel (where applicable) has worked well in the past. This traditional approach is at times challenged by the increasing sophistication and complexity of modern methods of analysis, design and construction. Design professionals who use these methods have a responsibility to provide code officials with adequate information to enable the necessary evaluations in an environment that is becoming increasingly competitive. But the code officials may lack the resources to adequately evaluate the conceptual basis and intent of these methods and their conformance with the construction codes that the officials are charged with enforcing. This section provides a mechanism for a jurisdiction to seek outside help for the review of such designs. A contract review is essentially a plan review conducted by a consultant or other third party. This review is considered a replacement for review by the jurisdiction. A peer review is a review on a higher, more theoretical and analytical level. This is not considered a replacement for plan review and will be discussed extensively within the User's Guide for this section.

The growing sophistication and complexity of design methods also leads to greater numbers of individual components, systems and processes that are interrelated to each other in ways that may not be well understood. The necessary understanding can often only be achieved by combining the resources of several consultants whose knowledge may be limited to that of individual components of particular systems. The methods often rely on materials, components and assemblies where acceptance may have been established by a series of tests in accordance with one or more nationally recognized consensus standards, but compatibility and relevance to the methods may not be fully established or understood. The code official is placed in the unenviable position of attempting to synthesize an often-vast amount of information in a coherent manner and in a limited amount of time in order to evaluate adequately these methods.

Peer review has the potential to enhance the quality and reliability of the design, review and construction of buildings, structures and facilities. It provides additional assurance of the completed project's performance by adding an independent and experienced voice to the process. The review would be performed by registered architects, engineers or special experts with knowledge and experience comparable to or exceeding those of the project's design professionals and comparable to the technical, conceptual and theoretical aspects of the project. It would not be a substitute for traditional plan review by code officials. Rather, it would be an additional review to test for the validity of the design and assist the code official in understanding how the design provides for minimum levels of public safety.

The concept of peer review and its importance has been recognized for numerous years by several organizations and professions, notably the structural engineering profession. Guidelines have been prepared and recommendations made to encourage the use of peer review for buildings where structural design uses analytical methods that are state-of-the-art or that are beyond the boundaries of what is currently acceptable by the structural engineering community. The reader is encouraged to study the following documents for more information:

- *Recommended Guidelines for the Practice of Structural Engineering in California*, 2020 Edition, 2000, Chapter 4.
- FEMA P-58-6, *Guidelines for Performance-Based Seismic Design of Buildings*, December 2018.
- *SEAOC Blue Book: Seismic Design Recommendations*, Structural Engineers Association of California, 2019.
- Section 1617A.14.1.41, Peer Review Requirements, 2022 *California Building Code*.

- The Society of Fire Protection Engineers also provides peer review guidelines that are available to set a baseline for such reviews. *Guidelines for Peer Review in the Fire Protection Design Process*, 2020.

The advent of performance-based design has increased the need for peer review of other professions involved in the design of buildings, structures and facilities. As innovative design methods and analytical procedures are developed for use in performance-based designs, peer review will become an increasingly important tool in assessing the use of these resources.

An important aspect of any process is an agreement on the meaning of the terms used to describe its components and concepts. The terms used for peer review vary in their meaning, depending on the point of view or focus of the organization or agency that prepares guidelines for the peer review's implementation. See Chapter 2 of the code for definitions of the following terms as they relate to peer review:

- Consultant.
- Contract review.
- Design documents.
- Peer review.
- Plan review.
- Quality assurance.
- Third-party review.

Buildings and structures designed in accordance with the provisions of prescriptive codes normally do not need to be considered candidates for peer review. However, virtually all project designs based on prescriptive code conformance have certain materials, components or systems whose acceptance is performance-based. The practice of structural engineering, for instance, is largely performance-based, incorporating the provision for rationality. Any system or method of construction must be based on a rational analysis in accordance with well-established principles of mechanics. The use of this and similar concepts will become more and more prevalent in other professions as performance-based design approaches are used. The reader should not assume that peer review is not warranted for projects designed to meet the requirements of prescriptive codes. A great example outside structural engineering is smoke control design, which is within the purview of the prescriptive building code. Similar to structural requirements in the prescriptive code, a rational analysis is required; in addition, there are elements such as the selection of a design fire that requires a higher level of review than a typical plan review can provide. Such designs often involve computational fluid dynamics models.

The code official decides when peer review will be required and the peer reviewer's scope of work. Generally, there are several conditions that prompt a code official to consider peer review for a project:

- One whose design is based on concepts, analytical methods and design procedures that go beyond the boundaries of what is currently thought to be acceptable by current code and professional standards. (See Sections 103.3.3 and 103.3.4 and Appendix C).
- One whose design is based on authoritative documents or design guides but is state-of-the-art and demands specialized knowledge to understand its underlying intent and objectives.
- One whose complexity or technical demands are beyond the resources normally available to the code official.
- One whose scope is such that review during the conceptual development of the design is considered critical to the eventual progression to construction documents.

The choice of a peer reviewer is critical to the effective use of peer review. Peer reviewers must be independent of the design professional, consultants, quality assurance personnel and contractors involved on the project. They must not have any vested interest in the project—financial, political, professional, personal or otherwise. They should also avoid and be free of all known or potential conflicts of interest. Individuals should not participate in a peer review process if they have any previous employment by, or financial interest in, any of the firms or companies involved in the design, review, inspection and construction of the project in its recent past. It is suggested that a peer reviewer serve for an established number of years, with that suggested term subject to being reduced based on the agreement of all parties involved.

Obviously, peer reviewers need to be qualified to perform the review. They should have a level of knowledge and experience at least comparable to that of the design professionals whose work they are reviewing. They should also have proven expertise in the design of projects of comparable complexity and theoretical demands. It is vital that the process of peer review be fair, objective and have a level of sophistication that matches that of the project's design and scope.

It will not be possible for peer review to be accomplished by a single person or firm for certain projects. The range of necessary expertise and experience is potentially vast. It is probable that separate peer reviewers with expertise in different professional disciplines will be needed, where applicable (for example, fire protection and structural engineering), and that should be encouraged. In such cases, it is important that the individual peer reviewers form a peer review team. A lead peer reviewer should be designated to coordinate all the elements of the review for the multiple disciplines and to ensure that all facets of the review are completed and submitted to the code official.

The code official should determine whether a particular individual is qualified to be a peer reviewer. The prescriptive codes generally afford the code official the discretion to require that plans, computations and specifications be prepared and designed by a registered design professional even if not required by state law. But the codes are understandably silent on whether the code official has the additional discretion to evaluate whether a design professional is qualified to perform a particular design. The code official usually relies on the rules and procedures of their state's professional boards to ensure the qualifications of design professionals.

Design professionals will initially achieve licensure or registration by a combination of experience and written examinations conducted by the state board. Then, their professional conduct is governed by a code of ethics. Often, professional societies also have codes of ethics that apply to their members.

However, there is a difference in the conduct expected of peer reviewers versus design professionals. The reviewers must have proven knowledge and experience that is greater than the minimum that is required to become licensed or registered. Also, they must have no vested interest in the project. The discretion to verify this should remain with the code official. But the burden could be substantially reduced if professional codes of ethics were amended to include rules of conduct for peer reviewers.

Peer review will normally be limited to a critical examination of the proposed conceptual and analytical concepts, objectives and criteria involved in a design, and such a review can and should extend through the duration of the project. Peer reviews are most effective when they begin at the conceptual phase of a project. They are not intended to be a detailed review for the purpose of verifying conformance with code requirements, performance or prescriptive. Code officials will usually opt for contract review when the resources or expertise to perform the plan review are not available to them. Peer review is not intended to be a replacement for the review process for a jurisdiction, normally called plan review. Peer review is actually in addition to the plan review. When code officials do not have the capabilities internally to review more complex designs, they must look to an outside source to conduct the design review. In this code such a review is termed a "contract review." A contract reviewer can be used as the peer reviewer for projects that are less complex.

The value of peer review is to provide feedback during the course of project development when the designers are making critical decisions. This can be of substantial benefit, since it may lead to the avoidance of significant errors at a time when they are more easily corrected. The scope of the project can also be reviewed at an early stage to ensure that minimum levels of public safety will be provided. A peer review should simplify the eventual plan review of the design documents, since many of the significant disputes and differences of opinion among the design professional, peer reviewer and code official will presumably have been resolved by then. The plan reviewers can then focus their energies on the detailed aspects of the design and not be overly burdened by the resolution of more global code issues that are best resolved during design development.

Peer review can be of substantial practical benefit to the owner. It has the potential to achieve savings in design and construction costs and to minimize delays in project schedules. Owners, developers, designers and contractors are encouraged to consider peer review whether or not it is required by the code official. The same is true of plan review by the peer reviewer in addition to that done by the code official. The benefits of both can be real, measurable and cost effective.

The code official has the final authority to decide the scope of work for a peer review. Ideally, this would be accomplished in concert with the owner, design professional and contractor. The code official would decide the minimum scope of work, but all affected participants would be encouraged to establish a scope consistent with the complexity and technical demands of the project and not be limited by any notion of "code minimums." The greater the peer reviewer's scope of work, the more reliable will be the design development and review of the project. There may, at some point, be diminishing returns, but the extent of peer review in this country has been too scant for overconfidence. It would be far better to err on the side of caution and stipulate more peer review. If conditions warrant, there is also the potential for the scope to be altered during design development upon prior mutual agreement of the owner and the code official.

The responsibility for design, inspection and construction of a building or structure that undergoes peer review remains entirely with those design professionals, inspectors and contractors involved in the project. The peer reviewer will not usually have any ability to prepare or change construction documents or have any role in inspection and construction. As stated above, peer review is intended to enhance the quality and reliability of the design, review, and construction of buildings and structures and provide additional assurance regarding the performance of the completed project.

A typical peer review consists of a series of reviews of the design criteria documents at previously agreed upon intervals. The peer reviewer should consider any or all of the following issues, while not being limited to them:

- Act as a coordinator for the code official in concert with the registered design professional in responsible charge.
- Verify that areas of the design that are performance based are adequately determined and clearly documented.
- Determine which portions of the design rely on the use of acceptable methods (for example, accepted standards, authoritative documents and design guides). Verify that the methods meet performance-based code requirements and prescriptive code requirements (where applicable).
- Of the remaining portions of the design that do not rely on the use of accepted standards, authoritative documents or design guides, verify what rationales, if any, are used to justify the design and whether their use is adequately substantiated.
- Verify that the design objectives, criteria, assumptions and concepts comply with performance-based code requirements and prescriptive code requirements where applicable and are clearly documented.
- Verify that the registered design professional in responsible charge has prepared an outline of design responsibilities that specifies the scope of services for each vendor, supplier, subconsultant and subcontractor who has the responsibility for design and/or installation of components or portions of the overall design. This should include rules of procedure for coordination of installation details, resolution of physical conflicts and timely submittal of drawings for review by the registered design professional in responsible charge, contractor, code official and other individuals, where warranted.
- Verify that all applicable special inspections, tests, observations by the architect/engineer, maintenance and operational requirements for future use of the building, and other applicable elements of the quality assurance program have been

adequately determined and clearly documented. The individual or firm who will administer the program should be identified to the code official.

- Verify that all portions of the construction documents whose design and review will be deferred (deferred submittals) until after issuance of the initial construction permits have been adequately determined and clearly documented.
- Verify that the documentation of deferred submittals includes all proprietary products that rely on performance-based specifications (for example, roofing assemblies, penetration firestop systems, exterior insulation and finish systems, concrete anchors and steel decking).
- Verify that the registered design professional in responsible charge has coordinated all design documents for consistency and compatibility before submittal to the code official for plan review.
- Verify that all applicable accepted standards, authoritative documents and design guides used in performance-based design are documented and are accompanied by explanations of how they were utilized in substantiating design solutions to demonstrate compliance.
- Verify that adequate documentation is contained in the design documents to enable independent review by the code official of the conceptual basis for construction or installation of all materials, components and assemblies whose design and/or review will be deferred until after issuance of the initial construction permits.

The peer reviewer should prepare a written report that describes the services performed, the opinion of conceptual review for compliance with performance codes, and recommendations and basis for acceptance or rejection.

See the Appendix D of this User's Guide for sample guidelines for peer reviewers based on Section 1617A.14.1.41 of the 2022 *California Building Code*, Peer Review Requirements. Note that the 2022 *California Building Code* is based on the 2021 *International Building Code*® (IBC®). The peer review requirements are an amendment to the IBC.

102.3.7 Permits and inspections

Many inspectors traditionally have used code provisions for the basis of approving inspections and have given a lower priority to approved plans. This level of flexibility is not possible with performance-based designs.

Building and related inspectors are responsible to conduct inspections, witness tests or receive test documentation to verify that construction complies with the approved design documents (plans and other designated documents). This is similar to the prescriptive code process. The code official may require additional inspections through approved third-party quality assurance inspectors to supplement the code official's staff capabilities based on the type of construction work involved. This practice has been used successfully by many code officials to add inspection expertise for continuous construction activities or where complex inspection procedures are needed to verify that the construction complies with provisions similar to Chapter 17 of the IBC. Several code officials have preapproved third-party inspection agencies based on the firms' qualifications and quality assurance practices. Third-party inspectors, code official's staff and testing agencies should be required to provide documentation for inspection and testing results that form the basis for the code official to authorize occupancy when all requirements have been met.

An audit or verification process is suggested when third-party inspectors or agencies are used to verify that required procedures, inspection, testing methods and report submittals appropriately document activities to meet the requirements of the code official. When the code official does not have the staff to provide these services, an inspection-oriented peer review can be substituted to verify that inspections are being conducted and reported in accordance with the IBC, Chapter 17.

Performance-based designs will often pursue the use of new technologies, products and elements of systems not previously approved by product testing and listing agencies. Code officials should expand current procedures and encourage agencies to develop improved programs for product labeling, field certification of products and quality assurance processes to verify that products meet the intended requirements of the performance code. This is a tremendous challenge for firms in the testing, measurement and product certification business. It will also be a challenge for code officials to verify performance in compliance with performance-based testing and standards.

Code officials and their staff should require that approved agencies with applicable national credentials and/or accreditation test, measure and report on the acceptance of products for meeting certain standards and conditions.

102.3.8 Project documentation

This section is intended to clarify the full scope of documentation required for a project, which might vary significantly based on the code official's discretion and the complexity of the project. Performance-based designs require a much higher level of documentation as discussed in the code and this User's Guide. Complete documentation that provides a clear record of objectives and decisions made to provide a safe building or facility, including associated operational, use and maintenance responsibilities, can be a substantial benefit to the code official's staff and the public after completion of a project.

Section 102.3.8.3 provides for the recording of the bounding conditions of a design as a deed restriction. This can be done by the conditional certificate of occupancy with attached bounding conditions recorded to provide public notice to future buyers, lenders, or any other persons, who may be reviewing legal conditions via a due diligence review. It is not intended that the operations and maintenance manual be included with the deed restriction; however, its effect and the responsibility for compliance should be noted in the bounding conditions.

102.3.9 Certificates

Certificate of occupancy and temporary certificate of occupancy requirements for a simple building where conditions do not require follow-up inspection, testing and report procedures are required by this code and applicable I-Codes. However, where the building or structure is more complex and requires annual or other periodic inspection, testing or reports by the owner, a certificate of occupancy with conditions (reference Section 102.3.9.1.3) may be required to verify that bounding conditions are met as a condition for continued occupancy. An operations and maintenance manual is recommended for more complex buildings, and compliance with the document should be required as a condition of continued occupancy. Most buildings and facilities, regardless of how they are designed, have ongoing requirements for inspection, testing and maintenance.

Failure of an owner to meet the attached written conditions of a certificate of occupancy would be the basis of enforcement to seek compliance or restrict the owner/occupant from occupying those portions of the building or facility affected by the conditions. Enforcement of these procedures for a community might require input from multiple code officials, such as building, fire and mechanical.

This code establishes a second certificate for use in conjunction with buildings and facilities called a certificate of compliance (Section 102.3.9.2). The certificate of compliance may be issued to regulate storage, processes or uses independent of the overall building, which may also have a certificate of occupancy. Since it is desirable to have a regulatory instrument governing a specific use or operation that may be acted upon independently, the certificate of compliance may be suspended, revoked or not renewed for a specific area, use or element within a building without having to revoke the entire certificate of occupancy.

102.3.10 Maintenance

As discussed in Section 102.3.5, the design professional is bound to predetermined intervals of maintenance where required to comply with the design documents. These items should be reviewed with the registered design professional in responsible charge, owner and code official as a condition of approving the design and documented in a manner by which future owners will know their responsibilities for continued maintenance.

Once a building or facility is constructed and approved, the owner is responsible for maintaining it in accordance with the approved operations and maintenance manual. Critical areas determined by the design professional may require inspection, testing and/or service at predetermined intervals to ensure continued compliance for the life of the building. When required by the code official, these findings and recommendations must be detailed in written form and recorded as a property deed restriction or condition. Such documents should include a specific notation stating that requirements can only be released through the approval of the code official. These ongoing requirements are similar to the intent of practices in several countries where the owner files verification documents annually with the code official to verify that a building or facility is being maintained in a safe manner. If the documents are not filed, the occupancy certificate is subject to revocation or nonrenewal.

The maintenance responsibilities of an owner of a performance-based design facility could require a higher level of accountability than if the facility were built to prescriptive construction codes. It is intended that the code official who enforces the building, residential, electrical, fire, mechanical and plumbing codes be part of the review process for buildings and facilities approved under this code and for enforcement of future maintenance provisions. The code official who enforces any maintenance aspects, whether related to fire or other ongoing requirements, must be adequately prepared for inspection of buildings and facilities that are in part constructed using a performance-based design. These inspections shall be in accordance with the operations and maintenance manual.

Where the operations and maintenance manual is included as part of the approved documentation for a project, the owner is required to comply with the requirements and to provide documentation in accordance with the applicable conditions. A certificate of occupancy with conditions must be met to maintain an active certificate of occupancy. Failure to comply with these conditions, including the operations and maintenance manual, may result in revocation or nonrenewal of the certificate of occupancy.

102.3.11 Remodeling, addition or change/approval of use

Where a building, structure or facility is proposed to be remodeled, renovated or to undergo a significant change of use, a qualified registered design professional is required to evaluate the existing construction documents for the proposed change and submit a written report of the findings and recommendations. The code official then should evaluate the written report, intended design or change of use to determine if the proposal adversely impacts the existing building or facility and to determine compliance with this code based on prior approved performance design documents. Based on the findings, written comments should be provided to the design professional indicating either approval of the change or the requirements for correcting any adverse findings. Once the report is approved in concept, the design professional can proceed with the design or initiate change of use, which may require a new certificate of occupancy or certificate of compliance to be issued.

Procedural steps for applying the performance code to an existing building are also included in the preface of the code. As with new buildings and facilities, if multiple design professionals are involved, a single registered design professional in responsible charge must coordinate the work.

102.3.12 Administration and enforcement

This section recognizes that the performance code is focused on the guidelines necessary for designs that contain a performance aspect. Therefore, the normal administrative procedures and requirements found in the prescriptive codes are still necessary in most projects. This section refers the code user to those codes for appropriate requirements.

102.3.13 Violations

This section was created in response to a concern that a mechanism for dealing with someone who does not comply with the code was necessary. Section 102.3.13 makes noncompliance unlawful, thereby creating an enforcement mechanism.

SECTION 103—ACCEPTABLE METHODS

General

The Acceptable Methods section provides acceptance criteria that determine the range of possible and credible acceptable methodologies and technical tools to verify that performance-code objectives are met. It is the design professional's responsibility to use accepted methods to demonstrate compliance with the performance code. There are three basic options to demonstrate compliance:

- Prescriptive approach.
- Performance approach.
- Combination of prescriptive and performance approaches.

Performance-based design

Acceptable methods include any method such as an engineering standard, engineering practice, engineering tool or computer model that has been accepted in a peer review process or has received positive evaluations in a consensus process among qualified engineers, educators and researchers, and that has been validated in its ability to generate outcomes consistent with those claimed by the developer where used in accordance with the appropriate documentation. Safety and reliability factors that are included or are required to be added should be explicitly stated and based on accepted engineering theory, engineering practice or statistics. Section 103 specifically requires the use of an "authoritative document" or "design guide," which are both defined in Chapter 2.

Designs based strictly upon prescriptive codes satisfy the performance objectives of this code without any additional analysis or verification. The prescriptive codes are considered "authoritative documents." Where undertaking a design that contains a performance-based component, the designer must take additional steps to demonstrate compliance with the objectives of the code. Performance designs require verification against performance criteria and specific documentation to support the designs. In addition to the use of a "design guide" or "authoritative document," the following issues need to be addressed for performance-based design:

- Analysis and rationale are based on authoritative documents or design guides.
- Objectives, functional statements and performance requirements of applicable topics are met.
- Documents are applied by individuals or groups qualified in the principles needed for a particular analysis.
- Documents are written in clear, unambiguous language.
- Terms are defined when deviating from a dictionary definition.
- Designs use appropriate factors of safety with regard to the associated uncertainty of performance and level of risk.
- Designs are based on statistically significant evidence.
- Scope of application relative to the performance requirements is clearly defined.
- Design professional clearly defines goals and overall objectives of the project and design-performance levels.
- Design professional clearly states tools and models to be used and provides documentation applicable to the design analysis and scenarios.

Verification of compliance

Undertaking a performance-based design within this performance code requires following certain procedures so that known and available methods are used to focus on the performance of buildings and their components or systems. This includes the ability to predict levels of performance and verification of compliance in advance through one or more of the following:

1. Verification by tests, computations or measurements.
    - 1.1. Prescribed tests for indisputable evidence to demonstrate performance compliance.
    - 1.2. Performance can be calculated in advance using a mathematical model or computer models based on acceptable parameters (for example, structural design, energy and fire protection design).
    - 1.3. Measurement techniques may be used to demonstrate that the design performance levels will be met through scaled testing or measuring intended results (for example, means of egress, measurement of number of people per minute, fire exposure, and active or passive smoke control scenarios).
2. Application of models or examples of known performance, based on models or examples of similar performance.
    - 2.1. Examples or models function by interpreting code requirements and provide guidelines for equivalent solutions.
    - 2.2. Documented use of code compliance evaluations may provide technical descriptions of building methods that will result in code compliance.

3. Application of other methods to demonstrate compliance.
   3.1. The registered design professional in responsible charge has the burden to substantiate code compliance via performance verification.
   3.2. Using a consultant who has documented expertise and can verify that the solution meets design objectives and code requirements.
   3.3. Certification by an expert.
   3.4. Using peer review procedures to verify a project is based on advancing technologies, testing and methods for determining the level of performance.

Application of standards

Numerous standards exist for testing and evaluating building materials and products. These tests are primarily focused on the application of the prescriptive code provisions. The standard used to evaluate the material or product must be appropriate for the application in which it is used and limitations of the results of those tests must be understood. The following factors should be considered in determining the applicability of a standard.

- The material or product to be evaluated must comply with all the limitations included in the scope of the standard.
- The standard must evaluate the material or product's performance as it relates to the particular manner in which it is to be installed and utilized.
- The standard must realistically evaluate the conditions to which the material is likely to be exposed during fire loading or the other applicable loading conditions being evaluated by the performance-based design.

This code has generally avoided the reference of specific standards. The intent of this code when it was originally written was that standards must simply comply with the parameters in Section 103. Performance standards are also unique and only a few are available as they pertain to building and fire codes. There are some standards that exist that are more performance based. Examples of performance-based design and construction standards include the following:

Elevators

- ASME A17.7/CSA B44.7-2007—*Performance-Based Safety Code for Elevators and Escalators* (note that this standard has been referenced in Chapter 30 of the IBC)

Structural fire engineering

- SFPE S.01-2011—*Engineering Standard on Calculating Fire Exposures to Structures*
- NFPA 557-2012—*Standard for Determination of Fire Loads for Use in Structural Fire Protection Design*
- ASCE 7, Appendix E—Performance-Based Design Procedures for Fire Effects on Structures

Structural design

- ASCE/SEI 7-22—*Minimum Design Loads and Associated Criteria for Buildings and Other Structures*

Some may consider ASCE 7 as prescriptive; however, it does contain a more performance-based approach than what the prescriptive codes and standards typically provide for fire. ASCE 7 is a key reference in the IBC and is referenced in Chapter 5 of this document. It is currently the only standard referenced in this code.

Authoritative documents and design guides

Chapter 2 provides definitions for authoritative documents and design guides. Some of the various sources are mentioned in the discussion on standards above. Standards typically fall under the category of authoritative documents. Some additional examples of both authoritative documents and design guides are provided below:

- FEMA P-58-2013—*Seismic Performance Assessment of Buildings, Methodology and Implementation*
- *SFPE Engineering Guide to Performance-Based Fire Protection (2007)* (currently being developed into an ANSI standard)
- *SFPE Guide to Fire Risk Assessment* (2023)
- *SFPE Engineering Guide to Very Tall Buildings* (2021)
- *SFPE Guide to Human Behavior in Fire* (2018)

CHAPTER

2 DEFINITIONS

SECTION 202—DEFINED TERMS

General

In the compilation of a code, each word has the potential to change the meaning of a code section. Further, terms can often have multiple meanings depending on the context in which they are being used. Therefore, it is necessary to maintain a section of definitions wherein the writers clarify the specific meaning of certain words in order to manage consistency throughout the document.

Additionally, many of the definitions found within this chapter reflect the specific needs of performance-based designs. Examples of such terms include "Bounding conditions," "Construction documents" and "Design documents."

The definitions within the *International Code Council Performance Code® for Buildings and Facilities* (ICCPC®) are primarily self-explanatory but several terms warrant further discussion.

Bounding conditions

One of the more important terms used in this document is "Bounding conditions," which establishes limitations on changes to the building or facility systems or components, maintenance and operation features that are identified as critical or necessary preventative features to provide a safe environment for occupants. "Bounding conditions" are established by the design professional and have concurrence of the code official for performance-based designs. Essentially, "Bounding conditions" establish the sensitivity of a design to change. Such conditions should be contained within the documentation and also attached to a conditional Certificate of Occupancy. This term is also used within the SFPE's *Code Official's Guide to Performance-Based Design Review*.

Facility

This term is used throughout the document and in particular as a term indicating that structures or areas other than just buildings are considered to be included in the application of the provisions of this code, such as tank farms and outdoor hazardous materials storage areas. This definition is based on the definition for "Facility" found in the *International Fire Code®* (IFC®), which includes buildings. However, there was a concern surrounding the potential misapplication of the term with regard to Section 602, Accessibility. Therefore, there was a need for a separate definition for facility that more closely correlates with the ICC A117.1 standard. A definition was also added for "Essential facilities." "Essential facilities" relates to the buildings, such as hospitals or shelters, that are needed after an event such as an earthquake or a hurricane. Therefore, there are now three definitions related to facilities, including definition of the term "Facility" as it applies generally to the code, the term "Facility" as it relates to accessibility, and finally the term "Essential facilities," as discussed above.

Authoritative documents and design guides

Using authoritative documents and design guides for analysis, design and justification for use with performance-based design requires a significant standard of care to verify applicability to the intent of the performance code.

AUTHORITATIVE DOCUMENTS and **DESIGN GUIDES** include technical references that are widely accepted and utilized by design professionals, professional groups and technical societies that are active in the design of buildings and their systems. These documents should pass through at least one of the following development processes:

1. Documents developed through open consensus processes conducted by recognized governmental bodies.
2. Documents developed through open processes but conducted by professional or technical societies, code or standards organizations or recognized governmental bodies.
3. Documents that have undergone peer review processes and have been published in professional journals, conference reports and recognized technical publications.

DESIGN GUIDES are developed by architectural professional organizations, engineering professional organizations and technical societies and are published as guidance for use in performance-based design. Standards of practice in performance-based design aid in the selection and application of engineering standards, computational methods and other forms of scientific and technical information that are applicable to the methodologies selected in the design approach. Such documents also add a level of consistency to the process.

AUTHORITATIVE DOCUMENTS are documents typically developed in forums as identified in Item 1 or 2. **DESIGN GUIDES** are documents typically developed in forums identified in Item 2 or 3. The listed processes are distinguished from a process that incorporates only a limited number of individuals' opinions. Such limited documents may include research papers, theses, product-specific manufacturer's guidance documents and other technical papers. Documents that are not considered AUTHORITATIVE DOCUMENTS or DESIGN GUIDES may be able to be used for a design when they comply with Appendix C for "individually substantiated designs." Because of the limited review of such approaches, Section 104.3.4 specifically requires a peer review of such methods.

See also the User's Guide to Section 103, Acceptable Methods.

Serious injury

A definition for "Serious injury" was necessary to differentiate between other, less severe injuries. This assists in the application and understanding of the levels of impact in Chapter 3.

CHAPTER

3 DESIGN PERFORMANCE LEVELS

This chapter is unique to the *International Code Council Performance Code® for Buildings and Facilities* (ICCPC®). It is intended to provide a framework to establish minimum levels to which buildings or facilities should perform when subjected to events such as fires and natural hazards. The minimums established by this chapter are based on the types of risks associated with the use of the building or facility, the intended function of the building or facility and the importance of the building or facility to a community. This information is then compared with the type and sizes of events that may affect the building or facility. As noted in the forward of this document, it is intended that this chapter provide a link between the policy makers and the designers. In many respects, this chapter is the performance code equivalent of the height and area requirements, occupancy classifications and related requirements.

SECTION 301—MINIMUM PERFORMANCE

Limitations on the extent of damage or impact on a building or facility are provided through design performance levels to which the structure must conform when subjected to events of various magnitudes. Determination of the required design performance level for a building or facility and for a specific magnitude of event is determined based on the performance group classification. Sections 302 through 305 describe the use and occupancy classification process, the assignment of each use group to a performance group and the relationship between design performance levels and magnitude of event.

SECTION 302—USE AND OCCUPANCY CLASSIFICATION

Section 302 defines use and occupancy classification as a means to categorize buildings, structures and portions of buildings and structures by their primary use, the characteristics of the persons using them, the level of risk assumed by persons using them during and after certain hazard events and their importance to the local community.

The definitions provided in Appendix A are based on the fundamental definitions provided in Chapters 3 and 4 of the *International Building Code®* (IBC®) for general use and occupancy classifications and special use occupancies. These use and occupancy classifications were modified in some cases to better categorize the use group in terms of occupant characteristics, risk and importance. In addition, the definitions have been modified to include specific assumptions regarding the primary uses of buildings or structures, characteristics of persons using buildings or structures within that use group, the level(s) of risk assumed by persons using buildings or structures within that use and occupancy classification during and after certain hazard events, and the importance and associated risk factors of the buildings or structures within that use group to the local community. The intent is to force a closer evaluation of who is at risk, how that person is at risk, and what the societal expectations are regarding the levels of safety and the necessary building and facility performance to address the risks. This appendix is provided as a means to relate the prescriptive and performance codes, but need not be used if other means of determining risk factors are acceptable to the designer and the code official. In general, the employment of the occupancy and use classifications from the IBC demonstrates that the concepts presented in Chapter 3 of this code are similar to those addressed in the IBC and provides a fairly solid starting point since many aspects of the importance of the building or structure, hazards and occupant characteristics have been implicitly included over the years within these classifications.

The following factors are important to consider in addition to the prescriptive use and occupancy classifications. Again, many of these factors (occupant density factors, for example) may already be implicitly addressed within the prescriptive classifications.

Nature of the hazard. The nature of the hazard, whether it is likely to originate internal or external to the structure and how it may impact the occupants, the structure and the contents must be addressed. These factors are important as different hazards present different risks (e.g., fire versus earthquake). For a fire hazard, the primary risk includes toxic gases and heat, but for earthquakes the primary risk includes falling debris and collapse. There may be different vulnerable populations for these risks, and the impacts are clearly different. Whether the hazard originates internally or externally to the structure could also be important for many reasons, including the number of people impacted simultaneously. For example, an earthquake impacts a large area simultaneously, but fire tends to affect a more isolated area, such as a single building or floor of a building. It is unclear whether or not the prescriptive code has taken this into account, but in a performance code approach these broader issues may become more apparent.

Number of occupants. The number of persons normally occupying, visiting, employed in or otherwise using the building, structure or portion of the building or structure must be taken into account. The larger the number of persons, the higher the potential for multiple life loss. Additionally, the number of persons may be relevant as large-loss events are generally perceived as more devastating than large numbers of low-loss events (e.g., 100 people dying in one plane crash in a year is often perceived as being worse than 50,000 people dying annually in individual automobile accidents). As with the nature of the hazard, it is unclear whether these perceptions are taken into account within the prescriptive code. Generally, the focus has been on one building at a time. If a large number of persons will be in one location, it is expected that those persons will be reasonably protected from whatever hazards might befall them. In general, protection strategies should be selected that aim to prevent multiple deaths and serious injury from occurring, with the tolerable number of deaths and serious injuries reflective of the hazard, the occupants and the use.

Length of occupancy. The length of time the building is normally occupied affects the risk characteristics of the occupants. This factor is intended to help address life-safety protection needs given such variations as having a structure occupied 24 hours a day (e.g., hospitals), during business hours only (e.g., offices) or rarely if at all (e.g., storage facilities). It also plays a role in hazard detection strategies. In some cases, structures will be occupied infrequently but have large numbers of people when occupied (e.g., a sports stadium). As with all of these risk factors, such combinations must be considered.

Sleeping characteristics. Hazard-induced risks are higher (from many hazards) when people are asleep. Reaction times are slower, and strategies such as faster notification times may be warranted.

Familiarity. This topic examines whether the building occupants and other users are expected to be familiar with the building layout and means of egress. If a hazard is such that people need to egress a structure quickly to avoid injury or death, unfamiliar surroundings can lead to confusion, especially if a) lighting is not available, b) people are disoriented because of the hazard or c) people are focused on trying to help others.

Vulnerability. This topic examines whether a significant percentage of the building occupants are, or are expected to be, members of vulnerable population groups such as infants, young children, elderly persons and persons with conditions or impairments that could affect their ability to a) make decisions, b) egress without the physical assistance of others or c) tolerate adverse conditions. No protection strategy can ensure freedom from risk. To help decide what level of risk is acceptable, many regulations target protecting large percentages of the most vulnerable or sensitive populations. For example, in acute care hospitals where care recipients cannot be moved, a protect-in-place strategy is often taken, with multiple levels of redundant protection. More specifically, identifying vulnerable populations can be useful in selecting performance criteria for assessing protection schemes against specific hazards. For example, if fire is the hazard of interest, and the elderly are the vulnerable population of interest, incapacitation or thermal impact levels can be determined such that some predetermined percentage of the vulnerable population can be expected to reach a place of safety without being overcome by the hazard impacts.

Relationships. This topic examines whether the building occupants and other users have familial or dependent relationships. Those people who are dependent upon others are clearly at an increased level of risk and are considered vulnerable populations. Those who are responsible for others require special attention, as they may place themselves at higher levels of risk in order to care for their dependents. In some locations, such as hospitals, protection schemes often account for this concern. In other locations, such as residences, protection schemes may not normally consider delays in evacuation that may be incurred as one family member searches for another, for example.

SECTION 303—PERFORMANCE GROUPS

Background

Section 303 determines the performance group to which each building or facility should be assigned. The designation of a performance group is intended to capture the importance of the building or facility, the types of risks associated with that building or facility and the intended function of the building or facility. Performance groups are determined using Table 303.1, which provides risk categories for buildings and other structures. This table is a duplication of Table 1604.5 in the IBC. This table was chosen since the assignment of a building or facility to a particular performance group based on risk category is a value judgment and is not technical in nature. This table has been discussed by a broader group of stakeholders over a number of years. This is similar to the use of occupancy and use group classifications from the prescriptive codes. The use and occupancy classifications for buildings are determined from the prescriptive IBC and address levels of safety versus hazard by requiring safety systems, such as fire alarm systems, automatic sprinkler systems and fire-resistance-rated construction as appropriate to the perceived hazard. However, these requirements are often public-policy-based decisions that are reactions to the perceived hazard, which may not directly correlate with the actual risk.

One characteristic of the table from Chapter 16 of the IBC is that it indicates that a natural hazard event, such as an earthquake, affects a building differently than a technological event, such as a fire. An earthquake affects a broad range of buildings and facilities at the same time, but a fire is usually limited to a single building or facility. A community is less likely to need extensive facilities for shelter after a fire as compared to an earthquake or perhaps a tropical storm. Also, the magnitude and frequency of a natural hazard cannot be controlled, whereas a technological event is a function of its surroundings. The issue of an event being a function of its surroundings is discussed in more detail under Section 305, Magnitudes of Event. It is worth mentioning that in recent years wildfires are impacting communities more broadly, similar to an earthquake or hurricane, and are linked more to a natural hazard event than a purely technological event.

Importance

The purpose for considering risk is that some specific buildings or other structures play critical roles in providing health, safety and welfare to communities, especially after hazard events. For these buildings or other structures, society demands a higher level of protection. There is an expectation that they remain functional after the event. In some cases this expectation is regardless of the magnitude of the event. In determining the importance to the community, based on societal health, safety and welfare objectives, the following were considered:

- The use or function of the building, structure or portion of the building or structure in providing a health-related service, such as hospitals. These facilities are expected to a) contain vulnerable populations who cannot be moved during a hazard event, and b) provide emergency medical services after a hazard event.

- The use or function of the building, structure or portion of the building or structure in providing a safety-related service, such as fire and police stations. These structures house essential safety service persons, equipment and communications, and are expected to be functional during and after hazard events.
- The use or function of the building, structure or portion of the building or structure in providing a societal-related service, such as the sole school with all of the community's children in attendance. The life-safety of the children and the significant role of a school to a community increase its level of importance to the community.
- The use or function of the building, structure or portion of the building or structure in providing a community welfare-related service, such as the primary employer for the community. The community may decide it would not be able to survive without it and may desire additional protection against any number of hazard events. The level of performance must also be balanced with cost effectiveness.
- The use or function of the building, structure or portion of the building or structure after a hazard event, such as a building used as an emergency shelter. Some structures, such as schools or places of religious worship, may be designated emergency shelters and are expected to function during and after hazard events.

Performance group designation

A performance group is a designation that identifies the required performance of a building or facility when subjected to a particular magnitude design event. The magnitude may be based on historical statistical data or on development of credible scenarios. A performance group is described by defining a set of maximum tolerable impacts (levels of performance) for a set of specified magnitudes of design events.

The main criterion for determining the performance group in which a particular building should be classified is to decide which use group or occupancy classification is appropriate for the particular building. Alternatively, one needs to consider the hazards and risks associated with a specific building or facility in conjunction with societal expectations regarding the level of safety. Following the use group classification approach, one uses Table 303.1 to determine the performance group based on the risk category. This table lists several facilities, but in many cases the performance group for facilities may need to be determined based on the risk factors discussed in Section 302 and relative hazards discussed in those use groups that are listed in Table 303.1.

Four performance groups have been established for this code:

Performance Group I. This performance group covers buildings or facilities, such as barns and utility sheds, where hazard-induced failure poses a low risk to human life. This group primarily includes utility-type buildings in which there is a low reasonable expectation of performance.

Performance Group II. This performance group is the minimum for most buildings.

Performance Group III. This performance group includes buildings and facilities with an increased level of societal benefit or importance or large occupant load. Examples include post-disaster command control centers, acute care hospitals and a school used as an emergency shelter.

Buildings and other structures that a) are equipped with a reliable means of limiting the area of impact resulting from an explosion or a release of highly toxic gas, and b) contain limited quantities of explosive materials or highly toxic gases can be classified under this performance group.

In hurricane-prone regions, buildings and other structures that contain toxic, explosive or other hazardous substances and that do not qualify as Performance Group IV structures shall be eligible for classification as Performance Group II structures for wind loads if these structures are operated in accordance with mandatory procedures that are acceptable to the code official and that effectively diminish the effects of wind on critical structural elements or that protect against harmful-substance release during and after hurricanes.

Performance Group IV. The highest performance group contains buildings or facilities that pose an unusually high risk. Such facilities may include nuclear facilities or explosive storage facilities. These buildings, facilities and classes of structures require increased levels of performance as they are expected to continue operations after a hazard. Their failure to do so could have a devastating effect within and/or outside the facility with any size incident. Certain businesses or facilities, such as semiconductor facilities, may voluntarily place themselves in this category because of the business interruption caused by a very small event.

Local government may increase the performance of any class of buildings (see the "Adoption" section in the Preface) if there are specific reasons. These reasons might include a situation in which the facility is the only employer, the only school or the only hospital. Likewise, should a building owner desire a higher level of performance for a specific building and design load, the level of performance may be increased during preliminary design. A worksheet that assists with determining the level of performance required is provided in Appendix B. Performance cannot be reduced below this level without approval by the appropriate authority. Also, adjustments may be necessary in certain use conditions based on, for example, a higher population of adults with disabilities than is normally expected in that particular use group.

Performance group application

Once the performance group is established, it is then applied to Table 303.3 to determine the tolerable level of damage and impact based on the performance group and the magnitude of event.

In the table, the horizontal axis contains the performance level (level of damage/impact), and the vertical axis contains a representation of the magnitude of event. As one reads from left to right along the horizontal axis, the performance groups increase, and thus, the allowable impact or damage decreases. As one reads from bottom to top along the vertical axis, the magnitude of event increases.

To use the table, the performance group classification that applies to the building or facility in question must be identified. One can then locate the appropriate magnitude of event and allowable impacts for the performance group.

Structures must be designed to the levels of performance and magnitudes of event indicated in every applicable square within Table 303.3. This can be illustrated by the following relationships.

Performance Group I. This means that the performance of the building or facility shall be such that:

1. Small magnitude events are permitted to result in, but not exceed, moderate impacts.
2. Medium magnitude events are permitted to result in, but not exceed, high impacts.
3. Large magnitude events are permitted to result in, but not exceed, severe impacts.
4. Very large magnitude events are permitted to result in, but not exceed, severe impacts.

Performance Group II. This means that the performance of the building or facility shall be such that:

1. Small magnitude events are permitted to result in, but not exceed, mild impacts.
2. Medium magnitude events are permitted to result in, but not exceed, moderate impacts.
3. Large magnitude events are permitted to result in, but not exceed, high impacts.
4. Very large magnitude events are permitted to result in, but not exceed, severe impacts.

Performance Group III. This means that the performance of the building or facility shall be such that:

1. Small and medium magnitude events are permitted to result in, but not exceed, mild impacts.
2. Large magnitude events are permitted to result in, but not exceed, moderate impacts.
3. Very large magnitude events are permitted to result in, but not exceed, high impacts.
4. Severe impacts are not permitted for any magnitude of event foreseen by and described within the code.

Performance Group IV. This means that the performance of the building or facility shall be such that:

1. Small, medium, and large magnitude events are permitted to result in, but not exceed, mild impacts.
2. Very large magnitude events are permitted to result in, but not exceed, moderate impacts (high and severe impacts are not permitted for any magnitude of event foreseen by and described within the code).

Note that this approach serves two functions. First, it provides a benchmark for design loads against which a building must perform in an acceptable manner. Second, it recognizes that there is always some likelihood of a small event growing larger (e.g., for a fire event), and that the losses associated with large events can be significant for some performance groups. If a community, a building owner or other stakeholder believes the expected loss to be unacceptable, a higher level of performance may be warranted.

SECTION 304—MAXIMUM LEVEL OF IMPACT OR DAMAGE TO BE TOLERATED

Section 304 of the code establishes how a building or structure is expected to perform in terms of tolerable limits of impact under varying load conditions, based on the determination of Table 303.3.

Tolerable limits of impact reflect various limit states of damage, injury or loss. The term "tolerable" is used to reflect the fact that absolute protection is not possible, and that some damage, injury or loss is currently tolerated in structures, especially after a hazard event.

Additionally, the phrase "provide high confidence" is included to describe the extent to which these damage limits can be achieved. This addresses the fact that performance design inherently entails significant uncertainty and variability both with regard to the magnitude of event as well as the capacity of the facility to resist such events. Prescriptive codes provide only a high expectation that the intended performance will be met. Also, with regard to many issues addressed by the prescriptive code, the expectation and capacity of building design and use is difficult to determine. The term "impact" is used as a broad descriptor of damage, injury or other types of loss. The four levels of performance are provided to "bound" the expected performance of buildings and facilities when subjected to various design loads. The intent is to describe this performance in terms of variables that can be measured or calculated.

Section 304 of the code is based heavily on concepts used in the Federal Emergency Management Agency (FEMA) 273, FEMA 274, the Vision 2000 report [copyright Structural Engineers Association of California (SEAOC) 1995], and the Performance-Based Seismic Engineering Guidelines—Part I (Draft 1, SEAOC Seismology PBE Ad hoc Committee, May 5, 1998). These documents established performance levels in terms of after-event damage states (impacts) and establish hazard levels in terms of magnitude of event, which may be deterministic or probabilistic descriptions of the magnitude of the event. In these documents, as in this code, design performance levels consider structural integrity, building operation, injury to people and damage to contents. The magnitude of event (design load) reflects an increasing level of the event magnitude (e.g., ground motion). The term "design load" was chosen for this code because both normal loads (e.g., dead loads) and hazard events (e.g., snow, flood, earthquake, fire) can be expressed as loads. This allows the format in Chapter 3 of this code to be used with all loads that a structure must resist.

Section 304.2 of the code defines the fundamental limiting states of tolerable impact to which a building should be designed and constructed and correlates with those in Table 303.3.

Note: Section 304.2 provides only a skeletal description of levels of performance, as the tolerable impacts (limit states) will vary based on the load. As such, details on additional or specific tolerable limits of impact are found in other appropriate sections of the code.

As specified in Section 304, there are four design performance levels defined in terms of tolerable limits of impact to the structure, its contents and its occupants: mild, moderate, high and severe. The language used reflects that although no amount of protection can guarantee complete prevention of damage, injury or death as a result of a hazard event (e.g., damage, injury or death could occur indirectly due to unknown conditions or reactions), some criteria for assessing compliance are required. The designer must translate this language into specific numerical criteria based on the specific situation and the supporting data used. This criteria must be approved by the code official. Such criteria could be formally assembled as part of the code as adopted where appropriate supporting data is available.

SECTION 305—MAGNITUDES OF EVENT

A large number of normal, natural hazard and technological hazard-based loads or events of various magnitudes can reasonably be expected to impact on a building or facility during its projected life span. These loads and events can vary across a broad spectrum, from seismic, wind, temperature and water on the natural hazard side, to fire, explosion, moisture, occupant safety and air quality hazards on the technological side. Normal loads and events can also vary broadly, from the myriad of live and dead loads associated with a structure to factors such as the potential for changes in soil conditions due to temperature and moisture variations. In order to evaluate the performance of a building or facility against these loads and events, a representative number of design loads needs to be considered and applied. (For simplification purposes, the term "design load" covers normal and hazard events as well.)

Design loads are characterized by four classes: small, medium, large and very large, indicating increasing magnitudes. Some design loads may be expressed as point values, whereas others may be expressed as distributions. As each type of load has unique characteristics, details are not provided in Chapter 3, but rather are provided in appropriate chapters of the code [e.g., Stability (Chapter 5), Fire Safety (Chapter 15) and Hazardous Materials (Chapter 19)] and are based on the Committee's understanding of current practice and limits on quantification.

In general, design loads may be defined, quantified and expressed deterministically or probabilistically. How these loads are expressed also varies by type. For example, a current approach to earthquake loads involves a probability of exceedance in a 50-year period. A very rare earthquake, or "very large" design load earthquake (in the parlance of this code), would be a very large magnitude event. This is a probabilistic approach. When designing for snow, however, the design load may be expressed in terms of a ground snow load, based on historical data, modified by exposure and risk factors. This is a deterministic approach. Fire loads may be expressed in such terms as heat release rate or mass (smoke) production rate and may have an associated time component (e.g., a 5 MW fire for 10 minutes). These would also be considered deterministic.

With regard to subjects like fire, the definition of design loads is dependent on the measurement of performance, which is in turn based on the use of the building or structure. In this code, performance is expressed in terms of tolerable impacts on buildings or structures, occupants and contents. Thus, a mild fire impact to contents of an office may be different than a mild impact in a fabrication area in a semiconductor facility. As a result, a small design load fire for an office may be different than a small design load fire for the fabrication area. Similarly, the mild impact to contents of an office may be different than a mild impact to an occupant of the office, and the resulting design load fires may be different as well. For earthquakes, the design loads are described generically in terms of mean recurrence intervals and are unrelated to the building.

When considering and developing design loads for a particular building or facility, it is imperative that the various design professionals associated with a project consider the range of possible events and how they may impact a building or facility beyond the earthquake and fire events previously discussed. A wide range of design situations and scenarios must be considered, including, for example, the possibility of changing soil conditions or the possibility of moisture accumulation. How the design loads may impact a building or facility and that different design loads may have different impacts must also be considered.

For example, soil expansion can create different magnitude of event levels and associated impacts for different systems or subsystems of a building; the impact on structural stability may be low, yet the architectural appearance may be impacted significantly. In addition, certain critical features may be significantly impacted. For example, the shifting of a foundation can lead to the inability to open an exit door or close a door in a fire or smoke barrier, or to an unacceptable impact on utilities such as the rupturing of water or gas supply lines into the building or facility. Likewise, moisture-induced expansion of building elements can also result in the inability to open an exit door or close a door in a fire or smoke barrier. These conditions can likely exceed design performance levels.

As discussed previously, the quantification of loads such as soil expansion or moisture accumulation can be probabilistic or deterministic. For any building or facility and its site, it should be possible to assess the likelihood of soil expansion, moisture build-up, temperature variation and other factors. One can then compare this assessment with the tolerable levels of impact defined earlier.

Section 305 of the code addresses the magnitudes of event that "can be reasonably expected to impact on a building." Terrorist activities, such as the 1995 Murrah Building bombing in Oklahoma City, the 1993 and 2001 attacks on the World Trade Center in New York City, and the 1998 embassy bombings in the Kenyan and Tazmanian capitals, heightened awareness in the building community to its role in managing the impact of such events. While prevention is generally considered a function of security, there are design considerations that can minimize the minimize the impact of terrorist attacks. Historically, the codes have not dealt with such extraordinary events, but this may change as the codes continue to evolve. A code such as the ICCPC provides a framework where such events could be considered, should the decision be made to design buildings or building features to address such events.

Given such variations in design loads and impacts as described previously, and in some cases the lack of readily available methodologies and data, the definition, quantification and expression of design loads is best accomplished by the appropriate professionals (e.g., structural, seismic and fire protection engineers) using the design performance levels established by this code. Also, given the broad spectrum of loads that may impact a building, it is imperative that all design professionals be involved in the process and that all realistic events or conditions that may impact a building or facility are considered in an appropriate manner. In the end, it is the responsibility of the design professionals to identify and evaluate an appropriate number of scenarios to validate the design analysis, the design details (e.g., system and component performance), material and product specification and ultimately, the material and product selection and installation in regard to the objectives, functional statements and performance requirements of the code.

Acceptable methods

Design performance levels are considered the performance criteria, and the magnitudes of event are considered the design loads and stresses. To increase the effectiveness of this code it would be ideal to create packages of technical performance criteria and design loads that fit within the context of these design performance levels and magnitude of events.

Due to the complexity of the issues, there will never be one single prescriptive solution for all designs. The I-Codes have been deemed to satisfy at least one of the acceptable methods for complying with the performance code. Essentially, buildings and facilities or portions of buildings and facilities that are designed and constructed in accordance with all applicable requirements of the IBC and IFC associated with the uses and occupancies listed in Chapters 3 and 4 shall be deemed to comply with the performance groups for that use group or occupancy. For example, a school designed and built to all applicable requirements in the IBC for an educational occupancy is deemed to comply with the performance group requirements for a building in the Educational Occupancy.

Though it is assumed that the IBC is deemed to comply with the design performance levels outlined in this code, the performance of buildings designed and constructed according to the IBC has not been specifically determined.

As noted, there is no single acceptable method to determine performance. Rather, a suite of acceptable methods (acceptable analytical tools and methods) is required to be applied to demonstrate that the design performance levels and magnitudes of event comply with the performance group requirements for the pertinent use groups or occupancy types. Examples of such acceptable methods related to fire include *SFPE Engineering Guide to Performance-Based Fire Protection (2007).* SFPE promulgates two structural fire engineering standards: SFPE S.01-2011, *Engineering Standard on Calculating Fire Exposures to Structures;* and S.02-07, *Calculation Methods to Predict the Thermal Performance of Structures and Fire Resistive Assemblies.* Complementary to these standards is Appendix E of ASCE 7 (2022), *Performance Based Design Procedures for Fire Effects on Structures.* ASCE also developed the *Structural Fire Engineering Manual of Practice (2018).*

Example:

Assume: A high school (Grades 9–12) with an attendance of approximately 400 students is to be built in Anytown, Mystate, USA.

Step 1. The first step in determining the requirements of this section (and of the ICCPC) would be to turn to Section 302 and/or Appendix A of the code to determine under which use group classification the school would fall. Clearly, the school would fall under Educational, with information provided as follows:

A103.1.3 Educational. A building, structure, or portion of a building or structure in which six or more persons, generally under the age of 18, gather for formal educational purposes.

1. It shall be assumed that occupants, visitors, and employees are awake, alert, and familiar with the building or structure.
2. Persons under the age of 10 will require assistance in exiting, and persons 10 years of age and older will predominantly be able to exit without assistance.

These assumptions reflect nominal characteristics of persons using an educational occupancy and provide the basis for such estimations as time to recognize an alarm, begin to exit, and find the way to a place of safety. Additional characteristics can be used if the information is available and supportable.

In this case, persons under 10 years of age will normally not be expected to be in the building. This would be acceptable grounds for modifying any assumptions regarding the level of assistance required for helping children out of the school in the event of an emergency. Additional guidance for these and other assumptions in the form of Functional Statements and Performance Requirements is found in various sections of the code.

3. Risks of injury and health assumed by occupants, visitors and employees during their use of the building or structure are predominantly involuntary.

This assumption reflects the fact that the people using educational spaces have limited responsibility for their own safety and are relying on the owners, managers, employees and insurers of the space to provide an adequate level of safety.

In this case, one might want to make additional assumptions about the use of athletic facilities and any additional risks voluntarily assumed by the student athletes. Assumptions about the need for protection against sick building syndrome and other health-related effects associated with the close proximity of large numbers of persons might also need to be considered.

Additional guidance for these assumptions in the form of Functional Statements and Performance Requirements is found in various sections of the code.

4. Public expectations regarding the protection afforded those occupying, visiting or working in an educational building, structure or portion thereof are high.

This reflects the expectation that spaces wherein large populations of children are gathered will be afforded a high level of protection to avoid catastrophic losses; i.e., a large loss of life in a single space is perceived as being worse than the loss of one or two lives in multiple, smaller events.

These assumptions and the design performance levels provide the basis for structural requirements. Additional guidance for these and other assumptions in the form of Functional Statements and Performance Requirements is found in various sections of the code.

If one wanted to determine the basis for the assumptions included in the use group descriptions, one could reference the appropriate section of Appendix A of the User's Guide.

Step 2. Next, one refers to Section 303, Performance Groups, to determine the appropriate performance group based on the risk category for educational use buildings. The first place to look is Table 303.1. From Table 303.1, it is determined that the performance group will be dependent upon whether there are more than 250 students expected to attend the Anytown High School. Because the expected attendance is 400, Anytown High School would be placed in Performance Group III. As discussed in Section 303, the performance groups are based on the risk category table that is duplicated from Chapter 16 of the IBC.

In addition to the risk factors discussed previously, the school may be considered important as a structure that may serve a necessary purpose in the event of an emergency. This information would come forth in the application of the work sheet in Appendix B.

Step 3. Now that the school is classified as Performance Group III, one then would go to Table 303.3 to determine the appropriate design performance level for the associated magnitudes of event to which the school is likely to be subjected. The first thing that should be noted is that Performance Group III allows only minimal impact for the medium magnitude of event for design purposes as well as the small magnitude of event. This reflects the assumptions in the determination of use, risk factors and importance that there are higher societal expectations for the level of protection provided in schools than there are for many buildings in the other use groups, such as a typical office building.

Step 4. At this point, one could choose to take the prescriptive approach and simply meet all the applicable requirements for an Educational Occupancy found in the IBC, *International Fire Code*® (IFC®) and other related I-Codes. Alternatively, one could choose to take a performance-based approach. A third option is a hybrid approach where the prescriptive codes and related standards are used for design and construction of most of the building and a performance approach is focused on a single aspect such as the structural design.

Step 5. If the performance-based approach is taken, the next step is to look at the descriptions of the tolerable impact for the appropriate design performance levels indicated in Table 303.3. These provide a qualitative description of the design performance levels required and can be used directly for a deterministic performance-based design approach or, in conjunction with the magnitude of event (load) found within Section 305, can be used for a probabilistic performance-based approach. Specific details on design load-related levels of performance are found in appropriate chapters (e.g., Chapter 5, Stability; Chapter 6, Fire Safety).

For example, for a medium-magnitude event, the design performance level for a building in the Educational Use Group is Mild Impact as stated in Section 304.2.1 of the code as follows:

304.2.1 Mild impact. The tolerable impacts of the design loads are assumed as follows:

304.2.1.1 Structural damage. There is no structural damage, and the building or facility is safe to occupy.

304.2.1.2 Nonstructural systems. Nonstructural systems needed for normal building or facility use and emergency operations are fully operational.

304.2.1.3 Occupant hazards. Injuries to building or facility occupants are minimal in numbers and minor in nature. There is a very low likelihood of single or multiple life loss.

304.2.1.4 Overall extent of damage. Damage to building or facility contents is minimal in extent and minor in cost.

304.2.1.5 Hazardous materials. Minimal hazardous materials are released to the environment.

For more specific details on magnitudes of design loads and design performance levels for specific hazards, one references the descriptions in the appropriate sections of the code (e.g., Chapter 5, Stability; Chapter 15, Fire Prevention and Impact Management). Similarly, to determine additional performance requirements that need to be met, the designer would reference the Functional Statements and Performance Requirements provided in Chapters 5 through 19 of the code.

Step 6. Given defined magnitude of event, design performance levels and commentary as discussed previously, a structure can be designed. In the case of the structural design, one would take the magnitude of event and design performance levels and translate them into loads and resistances. Guidance on translating the ground motion into loads can be found in acceptable solutions (e.g., prescriptive code, SEAOC Blue Book, ASCE 7, etc.) where a set of maps and formulas provide a set of loads, based on geophysical conditions, that the structural engineer can apply to the structural design process. Similarly, fire protection engineers would take the defined hazard levels, frequency and extent of growth for the design fire condition and, with the defined design performance levels (see Section 1502.2 for further information on design performance levels as they pertain to fire) and assumptions, design appropriate fire safety measures using acceptable methods. Note that Section 1502.2 makes a distinction between life-safety and property protection as it relates to fire. In other words, there is only one level of performance for life safety but multiple levels of property protection based on the performance group and occupancy or use of the building. Also, an upper limit on the level of property damage is placed on each performance group. For example, a building classified as Performance Group III would be limited to mild property damage for any magnitude fire.

CHAPTER 4

RELIABILITY AND DURABILITY

SECTION 401—RELIABILITY

This chapter addresses the importance of the reliability of individual protection systems and strategies, as well as the reliability of the interaction of these systems in achieving the design performance level for a particular building or facility addressed in Chapter 3. Reliability is a function of the many factors discussed below, including redundancy, maintenance, durability of materials, quality of installations and integrity of design. The discussion is primarily focused on fire safety systems and strategies but is intended to address other aspects of building design such as structural stability, mechanical systems and plumbing.

Systems reliability

Under prescriptive codes, the typical failure rates of systems are sometimes compensated for by requiring redundancy. One of the perceived advantages of a performance design is that it might allow the designer to minimize redundancy in order to achieve cost efficiency by increasing the reliability of the systems and/or strategies used to implement the design. In other words, increasing the number and effectiveness of the layers of protection provided may not necessarily increase the overall reliability of a design if probabilities of successful operation are not factored in. Therefore, a focus upon reliability should be established. Redundancy is only one, though important, way of achieving reliability. Reliability should be explicitly accounted for in the performance analysis. As part of this analysis it is important that factors affecting reliability over the life of the building be understood and addressed.

Operational reliability

Operational reliability is defined as the probability that a system or component will function as intended when called upon. A reliability of 100 percent means that the system will always work. Of course, because there is always a chance of something going wrong, 100-percent reliability can never be achieved.

Reliability analysis is a science used where the proper functioning of systems is crucial; the military, aircraft industry and nuclear power plant operators all apply reliability analysis to their systems. The reliability of a system is a composite of the reliability of its component parts, with a reliability of 1 being the same as 100 percent reliable. Mathematically, this can actually be calculated as 1 minus the sum of the probability of the failure rates of the component parts.

However, the failure of certain parts may not result in a failure of the system to perform within its intended range of performance. These are called noncritical parts and should not be included in the reliability calculation. This simplest approach to reliability analysis is called the parts count method.

However, the failure of several noncritical parts could lead to the failure of a critical part, but this mode of failure would not be addressed by a parts-count reliability analysis. Therefore, a more detailed analytical method called Failure Modes and Effects Criticality Analysis (FMECA) may be necessary. With FMECA, the failure modes of each component, as well as the probability of the occurrence of those failure modes, are evaluated as to the effect they will have on other components. For example, an electronic part that fails 75 percent of the time when it is open may have no detrimental effect, but if the electronic part fails 25 percent of the time when it is shorted, it may overstress a critical part and cause it to fail. These failure modes and probabilities are then incorporated into the overall reliability analysis, which can become quite complex, especially for a complex system.

Reliability of fire protection systems

These same concepts can be applied to fire protection systems or other critical safety systems in buildings incorporating a performance design. The greatest obstacle to conducting such an analysis is the lack of failure rate data on certain systems components. The major users of reliability analysis have kept detailed failure records on equipment for many years. These data do not exist for most systems found in buildings, although some data exists in surveys and in data kept by insurance inspectors. From these, estimates of systems reliability can be derived or approximated. In the future, manufacturers and service/maintenance companies will need to establish databases based on testing and field performance.

The most data available for fire protection systems and methods appear to be for sprinkler systems. Experience with commercial (NFPA 13) sprinkler systems indicates a fairly high operational reliability. Further, the data indicate that about half of the operational failures are attributed to some impairment in the water supply. For example, shut valves, clogged pipes, pump failures or problems with the municipal supply are the more common causes. Depending on where the problem is within the system, this could mean that only a few sprinklers might be affected or that the entire system might be impaired. This important distinction must be taken into account in the performance design analysis.

Data on the operational reliability of fire alarm systems suggest that they may not be as reliable as commercial sprinkler systems. But here, many of the failures appear to occur in individual detectors. Because detectors operate independently in a parallel wiring scheme (the exception being when one fails shorted, which results in a false alarm signal), problems with fire alarm systems tend to be more localized, with the remainder of the system continuing to operate normally. Addressable and intelligent technology increases reliability by more specifically targeting detector activation and by providing the ability of detectors to be adjustable for sensitivities to elements such as dust.

Although not often thought of in this way, fire-resistance-rated construction has reliability associated with it as well. Assemblies are furnace tested under ASTM E119 to assign a fire-resistance rating based on exposure to a standard time-temperature curve. The assemblies must be constructed in accordance with the tested design to ensure proper performance, although some assemblies are not very sensitive to construction errors. For example, masonry walls seldom crack or come apart to allow the passage of fire and hot gases (one of the E119 acceptance criteria), but most fail when the temperature on the unexposed side exceeds a specific value (another E119 acceptance criterion). Thus, errors in construction that do not affect heat transfer through the wall will likely not lead to failure. On the other hand, gypsum walls typically fail when the gypsum wallboard on the exposed side falls off, allowing fire to penetrate to the interior of the wall assembly and quickly through the entire assembly. Constructing one of these walls with too few or the wrong type of fasteners or improperly installed fasteners could lead to a wall that has less than the tested fire resistance.

Such problems appear to be rare because data from insurance inspections indicate a very high operational reliability for fire-resistance-rated construction comparable to that of commercial sprinklers. Clearly, the weak link in fire-resistive barriers is with intentional openings installed in rated assemblies to accommodate doors, windows and utility penetrations. The same insurance sources estimate that there is a 50-percent likelihood that a fire door in a rated wall opening will be blocked open or otherwise impaired and therefore negate the fire-resistance-rating performance of the entire wall assembly. The reliability of both an individual construction assembly and the overall system of compartmentation are important but difficult to measure.

Design and installation

The design and installation of fire protection features and systems and other building systems, such as refrigeration systems, must be conducted properly, or such features and systems will not be reliable. Especially with active systems such as fire alarm, sprinkler and smoke management, the devices selected must be appropriate to the hazards, and the installation must be correct. Many systems require a commissioning process that tests the full range of operation and sometimes includes third-party oversight, resulting in a certification. In some cases there are national bodies that certify competence, and many manufacturers offer training programs on the proper installation of their equipment.

Testing and maintenance

Testing and maintenance have a significant effect on the reliability of components and systems. Maintenance prevents failures by reducing wear and stopping problems before they start. Testing does not prevent failures but rather identifies failed components so they can be repaired before the system is needed. Testing must be done more frequently than the time between incidents, and repairs must be done promptly so that the system is working when it is needed. Maintenance must be done properly and at the required intervals so that detrimental effects are avoided. If not, the reliability of the system can be reduced significantly.

There are recognized standards for the testing methods and intervals for active fire protection features such as fire alarm systems (NFPA 72, Chapter 14) and fire sprinklers (NFPA 25). Maintenance required of components and systems is more individualized and is specified by the manufacturer. It is crucial to reliability that the maintenance and testing be performed as required by qualified personnel in order to avoid the introduction of problems by the very process used to avoid problems. Technicians performing testing and maintenance should be certified or at least working under the supervision of someone certified to work on the systems.

SECTION 402—DURABILITY

The objective of this section is to ensure that the building materials selected for a structure or facility are sufficiently durable or are repaired or replaced in a timely manner so that the performance objectives of this code are achieved and maintained throughout the life of the facility. The durability of a specific material, component or system should be appropriate for its use within the structure or facility and also consistent with its purpose in contributing to the desired level of building performance. The current codes indirectly address this issue by requiring certain types of materials. The durability of building elements, components and systems contributes to the overall reliability of the entire building as a system.

The selection of building materials and the protection, preservation and functionality of those materials should be such that the building will continue to satisfy the objectives of the code throughout its life. This may mean that in order to comply with the durability requirements, regular maintenance or replacement must occur. Performance objectives may impact the life-cycle cost of regular maintenance. Event magnitudes selected to meet tolerable impact limitations may influence the useful life of components and systems. For example, the design-performance objective for expansive soils might state a moderate level of tolerable damage to the foundation. But the performance objective for structural stability and means of egress might specify that only mild levels of damage can be tolerated. The design team needs to consider how various event scenarios influence or impact other performance objectives and tolerable limits of damage and how they in turn impact the durability of building components and systems. A design allowing for differential settlements might be well within the limits of moderate impact for the foundation; however, architectural components, HVAC equipment and floor surfaces could be rendered inoperable or unusable while their durability is adversely affected, negating their contribution to other performance objectives.

The performance of a building is dependent on the materials used in construction and the maintenance of those materials throughout the life of the building. It is anticipated that some materials, such as roofs, will need to be replaced, whereas other materials need only be maintained, such as the paint on an exterior wall. Use and exposure to physical stresses and environmental conditions also impact durability. These factors should be considered within the context of how materials are to be used in or on a building in conjunction with their intended function and their overall contribution to the building's performance.

The determination of whether a material for a specific element must last for the entire anticipated life of the building or can be one that is maintained, repaired or replaced as appropriate may well be dependent on the ability to access the element for inspection and maintenance and the importance of the element's contribution to the structural and fire- and life-safety performance of the building. For example, the paint on the exterior bearing wall of a building is easily inspected and maintained and also has a minimal effect on the structural and fire- and life-safety performance objectives during a fire event, but the fire protection encasing a steel column that is located within a framed wall is not as easy to inspect or maintain and is also essential to the structural integrity and fire- and life-safety of the structure during a fire event. Therefore, the paint could be durable for any length of time acceptable to the owner, but the fireproofing material must be durable for a much longer period of time.

The designer must include in the construction documents not only a description of the specific materials being used but also a description of the durability of each material ("a 20-year built-up composition roof, manufactured by the ABC Roofing Company," for example) or must specify the maintenance interval for each material (such as "exterior wood wall shall be painted every 10 years with an exterior grade, weather-resistant paint").

The replacement and maintenance schedules for building materials that are unusual may be required to be documented in the office of the authority having jurisdiction or in an affidavit signed by the owner and recorded so that future owners will be aware of the need for maintenance or replacement.

CHAPTER 5 STABILITY

SECTION 501—STRUCTURAL FORCES

General discussion

Section 501 provides the requirements for the structural design of buildings and other structures. This section specifies the forces that must be considered in the design of buildings or structures to achieve the required performance.

This section requires a structure to be designed for all the expected forces that the structure will be subjected to throughout its life. This is the same requirement found in Chapter 16 of the *International Building Code®* (IBC®).

The principles for the design of structures are the same regardless of whether the prescriptive or the performance code is applied. The methods for analysis will be the same for both approaches. The performance code gives the designer more flexibility in determining the expected forces and prescribes the performance of the structure when subjected to particular forces. The designer can look to the design performance level desired of the structure rather than simply applying a minimum solution.

Structural forces are related to the material chapters of the prescriptive code-including the design of masonry, wood, steel, concrete and aluminum-as well as to other portions of the code.

This document encourages building owners to become involved in the decision-making process that determines the design performance level of a building in relation to a specific event such as an earthquake. This code prescribes a minimum design performance level, based on the intended use of the building, but an owner may need to enhance the performance for different reasons. The current prescriptive approaches do not clearly state the performance level the code provides. Therefore, an owner is often not aware that they may not be getting the performance level desired from the building. The approach provided in this code, specifically in Chapter 3, is intended to better determine the appropriate performance level for a building or structure.

501.1 Objective

The objective statement for structural forces clarifies what is required with respect to the structural design of buildings and other structures. This statement has three parts, and each part has been written with specific intentions for structural design.

The first part of the statement, "to provide a desired level of structural performance," requires that buildings be designed to perform to acceptable levels. These levels may be similar for all structures subjected to a specific loading. For example, buildings must be designed so that all structural members are within allowable stress levels and acceptable deformations when subjected to expected dead and live loads. This would be a reasonable level of structural performance and would apply to all buildings.

However, there are times when different levels of structural performance are acceptable for a given event. For example, during a major earthquake (also known as a rare earthquake based on the frequency of occurrence), it would be reasonable to expect severe damage to a single-family dwelling, moderate damage to a school and minor damage to a hospital. In this example, these would all be reasonable levels of structural performance. In no event would collapse be a reasonable performance level.

The second part of the statement, "when structures are subjected to the loads that are expected," requires buildings and other structures to be designed for generally accepted loads and load combinations. For example, the IBC specifies the live load for various types of occupancies. These would be examples of generally accepted loads. The design of a concrete floor requires that the calculated floor dead load be added to the specified live load along with other applicable loads for partitions, mechanical equipment, etc., and that the floor be designed for the combination of all these loads. This is an example of a generally accepted load combination.

The last statement, "during construction or alteration and throughout the intended life," requires that buildings or other structures perform not only during their construction and subsequent alteration but also that the durability of the structure be sustained throughout its life.

The building or structure shall perform as intended under normal conditions. Performance levels deemed appropriate must be defined by the following limiting states of tolerable damage: structural and nonstructural performance levels shall be applicable to natural, technological and fire hazard events reasonably expected to impact the structural and nonstructural systems during the projected life of the building or structure.

> **Mild impact.** During and after a hazard event, basic vertical and lateral force-resisting systems of the building are expected to retain their entire prehazard event strength and stiffness. Minor structural damage that occurs as a result of a hazard event shall not delay reoccupancy.
>
> During and after a hazard event, nonstructural systems required for normal building use, including lighting, glazing, plumbing, HVAC and computer systems, must remain fully operational, although minor cleanup and repair of some items may be required. Basic access and life-safety systems, including doors, stairways, elevators, emergency lighting, fire alarms and suppression systems, shall be fully operational. Large or heavy items that pose a falling hazard, including parapets, cladding panels, heavy

plaster ceilings, suspended ceilings, glazing systems, lighting fixtures and storage racks, shall be designed to prevent damage or failure of the items from excessive movement.

Moderate impact. During and after a hazard event, basic vertical and lateral force-resisting systems of the building are expected to retain nearly all their prehazard event strength and stiffness. Moderate structural damage may occur as a result of the hazard event and will delay reoccupancy; however, the structural damage should not be so extensive as to prevent repair or rehabilitation.

During and after a hazard event, nonstructural systems required for normal building use, including lighting, glazing, plumbing, HVAC and computer systems, shall remain significantly functional, although cleanup and repair of some items may be required. Basic access and life-safety systems including doors, stairways, elevators, emergency lighting, fire alarms and suppression systems shall remain fully operational.

High impact. During and after a hazard event, structural elements and components are expected to have significant damage. However, the building or structure shall be designed such that hazards from large falling debris, either inside or outside the building, are prevented. The amount of structural damage shall not be such that repair of the structure is not possible; however, significant delays in reoccupancy or a decision not to repair the damage may result.

During and after a hazard event, nonstructural systems required for normal building use, including lighting, glazing, plumbing, HVAC and computer systems, may be significantly damaged, and their functions may be limited. Egress routes within the building may be impaired by lightweight debris, and the HVAC, plumbing and fire safety systems may be damaged, resulting in loss of function.

Severe impact. During and after a hazard event, significant degradation in the stiffness and strength of the lateral-force-resisting system, large permanent deformation of the structure and, to a more limited extent, degradation in vertical-load-carrying capacity must be expected. However, all significant components of the gravity-load-resisting system must continue to carry their gravity load demands. It shall be expected that the structure may not be technically practical to repair and is not safe for reoccupancy, as additional hazard event activity, even at a reduced level, could induce collapse.

During and after a hazard event, nonstructural systems required for normal building use, including lighting, plumbing, glazing, HVAC and computer systems may be completely nonfunctional. Egress routes within the building may be impaired by debris, and the HVAC, plumbing and fire safety systems may be significantly damaged, resulting in loss of function. The building or structure shall be designed to avoid failures that could injure large numbers of persons, either inside or outside the building or structure. Significant failure of large or heavy items such as parapets, cladding panels, heavy plaster ceilings, suspended ceilings, glazing systems, lighting fixtures or storage racks may pose a falling hazard.

501.2 Functional statements

The first statement specifies that the structure should be designed to provide a reasonable level of structural performance to protect the occupants from injury. Because the needs of the occupants vary, differing design performance levels would be required for critical occupancies such as hospitals and emergency rescue facilities, as well as for high occupancy buildings such as large theaters and auditoriums. A specific reference is made to Chapter 3, which provides guidance in determining the design performance level.

The second statement indicates that the structure must be designed and constructed to achieve acceptable performance to protect property, both on-site as well as adjacent to the site. The property (or "amenity," as used in this section) would not only be the structure itself but would also include its contents. Again, depending on the needs of the owner, occupants and community, this performance level may vary.

501.3 Performance requirements

The performance requirements are specified in seven sections. The first section requires that structures and portions of structures remain stable and not collapse during construction or alteration through the intended life of the structure. This section requires that structures be designed to address hazards during construction and alterations and that the materials used are durable and maintained throughout the life of the structure.

The second section recognizes that structures, from time to time, experience minor damage from a variety of hazards such as fires, small earthquakes, or overstresses due to concentrated loads. This section requires the structure as a whole to be capable of absorbing these local damage areas without causing major damage to the entire structure. For example, this may require that buildings be provided with more than one line of resistance in each direction for lateral loads.

The third section requires that the deformation of the structure from design loads be within tolerable limits. For example, under dead and live loads, the floor of a building should not vibrate or deflect so as to cause discomfort to the tenants. Also, under seismic loads, the structure's drift should be controlled so that it does not impact adjacent structures.

The fourth section specifies the forces that the structure must be designed to resist. A discussion of these loads is addressed as follows:

1., 2., 3. The design professional must evaluate all loads and combinations of loads as is accepted standard practice. For dead, live, impact and other loads not specifically commented on in Items 4 through 14, the designer must evaluate best current practice as recognized in authoritative documents. In all cases, the design engineer must demonstrate that the magnitude of events is appropriate for the performance level and magnitude of damage to be tolerated.

4. Explosion hazards can be described in terms of exceeding a defined energy release within a building, structure or portion of a building or structure. Pressure loads can also be used to define explosion hazards. In all cases, the design engineer must demonstrate that the magnitude of events is appropriate for the performance level and magnitude of damage to be tolerated.
5. Soil and hydrostatic loads can play a critical role in the performance of structures. These loads will be site specific and are heavily dependent on the type of soil and location relative to water sources.
6. Flood hazards are described in terms of the mean return period for the 1-percent annual chance flood event (100 years) and the 0.2-percent annual chance flood event (500 years). For many locations, the land adjacent to bodies of water that may be affected by floods of one or both of these mean return periods is shown on the Flood Insurance Rate Maps prepared by the Federal Emergency Management Agency (FEMA). The area that is expected to be inundated by the 1-percent annual chance flood is commonly known as the "A Zone" (the "V Zone" along some open coastlines). Where such maps do not show the presence of a flood hazard adjacent to a body of water and for flood hazards with mean return periods greater than the 100- or 500-year flood events, such events are to be determined on a site-specific basis. Standard methodologies are to be used to make the determinations and may include application of rainfall-runoff models, hydraulic models, storm surge models and evaluation of historic records of storm events and flooding. Some areas shown on FEMA's maps not subject to flooding by the 1- or 0.2-percent annual chance floods may be subject to flooding by lower probability, high consequence events (i.e., large, and very large), such as extreme storm surge flooding, levee failure, dam failure or tsunami flooding. ASCE 24, Flood Resistant Design and Construction, covers standard methodologies to account for flood hazards.
7. Wind hazards are described in terms of the mean return period of a defined magnitude of wind speed (3-second gust) in defined geographic areas (zones). The authoritative document, ASCE 7 and Commentary, was used as the primary reference for the mean return periods.
8. If there are other forces that are expected to affect the structural performance of the structure, the designer must also design for those other forces and consequences. Such forces can include, but are not limited to, windborne debris impact loads or hail impact loads and shall be accounted for in the design of structures to achieve the desired performance level.
9. Snow hazards are described in terms of the mean return period of a defined magnitude of surface snow precipitation in defined geographic areas (zones). The authoritative document, ASCE 7 and Commentary, was used as the primary reference for the mean return periods. Drifting, unbalanced loads, sliding snow loads and ice damming must also be addressed as, these loads can significantly affect the performance of a structure.
10. Rain loads are provided in terms of mean return intervals for both primary and secondary drainage capacity requirements.
11. Seismic hazards are described in terms of the mean return period of a defined magnitude of seismic-induced ground motion in defined geographic areas (zones). See the following section for more detailed information.

    The hazard for small earthquakes is referenced in the PEER Tall Building Initiative, Version 2 (PEER, 2017) and the Los Angeles Tall Building Structural Design Council document (2018). While these documents were drafted for tall buildings, their procedures have been used for design of other than tall buildings.

    Starting with the 2010 edition of ASCE 7, the Maximum Considered Earthquake definition changed to reflect a risk-targeted probabilistic hazard, as opposed to the uniform hazard defined by the 2,475-year return period. The risk-targeted probabilistic hazard for Large and Very Large earthquakes varies in return period across the country. Additionally, the ASCE 7-16 deterministic cap has changed from 150% of the mean shaking intensity of the characteristic earthquake to a mean plus one standard deviation shaking intensity from the characteristic earthquake, which is approximately 180% of the mean.

History of the seismic requirements in the ICCPC.

The initial ICCPC seismic requirements were based on the available resources at that time. The following discussion gives background as to what resources were considered.

In the 1996 *Recommended Lateral Force Requirements and Commentary* [Structural Engineering Association of California (SEAOC) Blue Book], Appendix B: Conceptual Framework for Performance-Based Seismic Design (Vision 2000), the design loads were specified by a series of four earthquake design levels (events) and were expressed as a corresponding set of probabilistic earthquake ground motions (p. 396).

"Recurrence interval" was comparable to "mean return period," and "frequent" through "very rare" events were comparable respectively to "small" through "very large" events in the performance code. The rare event was also specified as the design level earthquake ground motion (Commentary to 1996 SEAOC Blue Book, p. 97).

Event	Probability of Exceedance	Recurrence Interval
Frequent	50% in 30 years	43 years
Occasional	50% in 50 years	72 years
Rare	10% in 50 years	475 years
Very rare	10% in 100 years	970 years

In the 1998 *Performance-Based Seismic Engineering Guidelines, Part I—Strength Design Adaptation, Draft No. 1* (SEAOC Seismology PBE Ad hoc Committee, dated 5/98), the design load was specified the same as in Appendix B of the 1996 SEAOC Blue Book, except the probabilistic ground motions for the frequent and very rare events were revised. The recurrence interval for the frequent event was decreased. The revision was done partly to express the interval in terms of the same 50-year probability of exceedance as the occasional and rare events.

The very rare event was no longer specified probabilistically but deterministically as approximately 150 percent of the rare event. In the Western United States, this typically corresponds to a mean recurrence interval of 2,000+ years. This change was made partly because of the use of the very rare event design load for buildings near active faults. SEAOC specified a very rare event as 150 percent of a rare event.

Event	Probability of Exceedance	Recurrence Interval
Frequent	87% in 50 years	25 years
Occasional	50% in 50 years	72 years
Rare	10% in 50 years	475 years
Very rare	Not applicable	Not applicable

In the 1997 *NEHRP Recommended Provisions for Seismic Regulations for New Buildings and Other Structures* (FEMA 302 and 303, dated 2/97), the design load was specified at the very rare event level. The other event levels were not mentioned. The very rare event was expressed as the lesser of the probabilistic and deterministic maximum earthquake ground motions. The probabilistic level was that which would be developed with a 2-percent probability of being exceeded in 50 years (2 percent in 50 years), equivalent to an approximate mean recurrence interval of 2,500 years. The deterministic level was specified as 150 percent of the median 5 percent damped spectral response accelerations at all periods resulting from characteristic earthquakes on any known active fault within a particular region. The spectral response accelerations were mapped and included in the FEMA document.

The 1998 FEMA-273 NEHRP *Guidelines for the Seismic Rehabilitation of Buildings* used the following:

Event	Probability of Exceedance	Recurrence Interval
Medium	50% in 50 years	72 years
Large	10% in 50 years	474 years
Very large	2% in 50 years	2,475 years

A small event comparable to the frequent event specified in the 1998 *Performance-Based Seismic Guidelines, Part I* was not included. Also, an event between the medium and large events was included but was not found in any of the above references.

Based on these documents, the performance code initially used a combination of the 1998 SEAOC and FEMA 273 data as follows:

Event	Probability of Exceedance	Recurrence Interval
Small	87% in 50 years	25 years
Medium	50% in 50 years	72 years
Large	10% in 50 years	474 years
Very large	2% in 50 years	2,475 years

The language "Large" and "Very large" in the Event column originates from FEMA-302, *NEHRP Recommended Provisions for the Seismic Regulation of Buildings and Other Structures.* These particular provisions were also the basis for the IBC requirements. More specifically, in FEMA 302 and the IBC, Maximum Considered Earthquake (MCE) ground-shaking demands (similar to the very large demands in this code) in near-fault regions such as coastal California or the New Madrid region of Missouri were taken as the lesser of either probabilistically determined shaking or deterministic shaking based on the largest earthquake likely to occur in the region. Also, FEMA 302 and the IBC Design (DBE) ground-shaking demands (similar to large loading) were taken as demands with two-thirds the intensity of motion as MCE (Very large) demands.

12. Ice hazards are described in terms of the mean recurrence interval of a defined magnitude of surface icing in defined geographic areas (zones). The authoritative document, ASCE 7 and Commentary, was used as the primary reference for the mean return periods.
13. For hail loads see Item 8.
14. Thermal loads refer to hazards such as fire. This is a unique load to be included as it has historically been dealt with through prescriptive fire resistance and has not been directly referred to as a load. Note that Chapter 15 of this code addresses the need for structures to withstand the effects of fire. See the performance requirements in Section 1502.3.11. Thermal loads could also refer to other heat sources such as thermal loads from the placement of the building.

Section 501.3.5 recognizes the need to address uncertainties involved with design, construction, building use and material properties. Target reliabilities must be addressed in accordance with ASCE 7.

Section 501.3.6 requires that alterations and demolitions be done in a safe manner to avoid injury to the workers on the site and the public adjacent to the site. Section 703 also addresses construction and demolition hazards.

Section 501.3.7 requires that the grading of sites be done in a safe manner so as to prevent damage to the adjacent property.

Acceptable methods

There are several acceptable solutions available through current codes and standards. Chapter 16 of the IBC and ASCE 7 are already referenced in this chapter in terms of structural loads and target reliabilities. Other key chapters of the IBC that can serve as part of the acceptable solutions include Chapters 19 through 23 for specific material types. Those chapters also pull in material specific design standards. IBC Chapter 17 addresses target reliabilities, in part, through the use of special inspections. Chapter 18 addresses geotechnical and structural conditions focused on foundations. Chapter 33 of the IBC addresses site work and protection of adjoining properties, focused on construction and demolition.

The *International Residential Code* serves as a prescriptive solution for one- and two-family dwellings.

Existing buildings can be addressed prescriptively through the *International Existing Building Code*. This code focuses on standards such as ASCE 41, which provides a more performance-based approach to address seismic performance upgrades in structures.

CHAPTER

6 PEDESTRIAN CIRCULATION

SECTION 601—MEANS OF EGRESS

These provisions provide guidance by which egress systems for buildings and facilities are designed, evaluated and maintained. The current prescriptive codes dictate a solution of standardized elements based on narrowly defined minimums and maximums. The prescriptive methodology is successful in providing safe buildings, but it is found to be a hindrance in many cases in the design of new buildings because a single solution-set usually does not fit perfectly for every situation. The prescriptive approach tends to focus code users on specific numbers and therefore causes users to lose sight of the intent.

The general concepts of egress have not changed over time, nor have they changed as a result of the creation of performance codes. Safe egress continues to involve exiting the building or facility safely or relocating to a safe place, i.e., an area of refuge. A multitude of occupancy characteristics influence the decisions made on egress, ranging from an occupancy consisting of completely ambulatory, self-preservation-capable persons to an occupancy consisting of disabled, infirm or incarcerated persons.

All the elements found in the current prescriptive codes were considered in this section. The terms formerly used for these elements are not necessarily found in the text. Generic terms make it easier to expand the scope as may be necessary in the future when thinking in a true performance manner. The authors of these provisions were careful to avoid existing terminology and thereby avoided preconceived notions based on past definitions. As the egress provisions are revised in the *International Building Code®* (IBC®), Section 601 may not need revision as the overall objectives, functional statements and performance requirements will not change. For instance, the concept of using elevators for egress was added to the 2009 IBC as an alternative to an additional stairway in very tall high-rise buildings. The *International Code Council Performance Code® for Buildings and Facilities* (ICCPC®) has always been able to facilitate such design strategies as long as the design could demonstrate that it met the design performance levels, objectives, functional statements and performance requirements.

An effective system of egress is interdependent with provisions for accessibility, fall prevention, number of occupants, level of risk and building safety systems. Each of these issues has to be factored into acceptable solutions for egress.

Public discussion of these performance provisions has focused on the use and lack of use of terms defined in the prescriptive codes. In the performance section on egress, the term "safe place" is used and the terms "public way" and "area of refuge" are not. The term "safe place" is universal, as it may refer to an exterior or interior location. Additionally, it gives the building designer the information needed that people must be conducted to a safe place.

601.1 Objective

The objective conveys the ultimate goal of the chapter. The current prescriptive codes do not clearly state an objective, and the IBC and *International Fire Code®* (IFC®) commentary focus on the hazards related to fire. This code conspicuously avoids the term "fire" and focuses on the goal. There are other emergencies that may necessitate the evacuation of a building's occupants.

These provisions apply to all building occupants, including those with disabilities. The egress provisions do not cover rescue operations by emergency responders, which are specifically dealt with in Chapter 17, Emergency Access and Facilities.

601.2 Functional statement

The functional statements indicate that the building or facility must be designed, constructed and maintained to allow occupants to egress according to the design performance levels of Chapter 3. Section 1502 was revised fairly early in the development process to reflect a single performance level for life safety. A similar revision was not made to this section. Means of egress may need to be available for other incidents beyond fire, such as a hazardous materials release or an earthquake. See discussion regarding Section 1502.2 as to the reasoning for a single performance level for life safety with regard to fire.

601.3 Performance requirements

These provisions apply to both new and existing buildings and facilities where appropriate. The performance requirements are fairly intuitive and relate back to Chapter 10 of the IBC and IFC in intent. Some of the key issues are as follows:

- Travel distances.
- Number of occupants.
- Occupant characteristics.
- Identification and illumination.
- Safe place (area of refuge).
- Unobstructed path.

In a performance code, numerical values will not initially be provided for travel distance or occupant load. Likewise, terminology such as "exit passageway" will not be used. A performance-based egress solution will deal with the evaluation of the hazard and the available egress time for the occupants to avoid interfacing with the hazard. Some of the prescriptive egress requirements such as exit sign location and illumination, may remain prescriptive in most performance designs.

Section 601.3.6 clearly points out that regardless of when a building or facility was constructed, a suitable level of performance is still required. This aspect relates to the intent of the egress provisions in the prescriptive IFC. One of the traditional functions of fire codes has been to ensure that the egress system of a building or facility continues to function over time.

In many cases, it is difficult to truly understand the hazards that exist within an existing building, and it is even more difficult to understand which changes can be the most effective. Through an analysis, the designer and the code official may find, for example, that the width of the exits is much more important than the travel distance.

Also, as noted, egress is interdependent with other aspects of building functions and characteristics such as fall prevention, accessibility, hazards and characteristics of the occupants. Therefore, Section 601.3.5, relating to appropriate walking surfaces and facilities to avoid falls and injuries, was included. If surfaces and related features such as ceiling clearances are addressed for egress during an emergency, then falls and injuries related to falls should be the same or lower during nonemergency daily use.

SECTION 602—ACCESSIBILITY

Accessibility requirements provide disabled persons with reasonable access to and use of buildings to an extent consistent with that to which people without disabilities are able to access and use buildings. This is consistent with the intent of federal statutes enacted to protect the rights of people with disabilities.

This intent is implicit in all provisions of this code, whether or not the intent is specifically stated in each section. The goal of the performance code is to allow freedom of design while providing an acceptable level of accessibility throughout a building. Therefore, it is incumbent upon the user of the code to base decisions on reasonable criteria with the intent of providing equal access. Since this topic is focused on addressing civil rights and can have legal implications, designs should comply with IBC Chapter 11, ICC A117.1, federal regulations promulgated under the Americans with Disabilities Act and the Fair Housing Amendment Act, or other regulations promulgated under state disability rights statutes. This section is intended to lay out the intent of the accessibility provisions.

A specific effort was made to make this section an all-encompassing section that requires accessibility to be evaluated in each part of the code where applicable. This would include accessible egress.

SECTION 603—TRANSPORTATION EQUIPMENT

This section provides general safety guidelines for installation of elevators, dumbwaiters and escalators inside or outside buildings. These provisions include use during normal operations, use by emergency responders during emergency operations and use by maintenance personnel during activities associated with adjusting, servicing and inspecting elevators. The provisions focus on the expected loads, emergency recall for medical emergencies, firefighting and rescue operations and safety of maintenance personnel.

Other provisions not directly associated with elevators, dumbwaiters or escalator equipment, such as the protection of hoistway enclosures and electrical equipment, are covered by other provisions of this code such as management of fire impact.

As noted in the discussion within Section 601, the use of elevators during emergencies has evolved based on the findings in the World Trade Center study and also related to the evolving needs of society in very tall buildings. More specifically, egress based only upon the use of stairs in very tall buildings may not be practical, especially when an event would require full building evacuation. Occupants may have physical limitations preventing them from using the stairs or from walking down many flights of stairs. This includes wheelchair-bound occupants and occupants impacted by other physical limitations such as asthma, joint problems or pregnancy. The current objectives, functional statements and performance requirements accommodate the use of elevators for egress along with the egress requirements in Section 601. The concept of using elevators for egress had been formalized under certain conditions in the IBC. It should be noted that the IBC references ASME A17.7 as a viable design alternative for elevators (see Section 3008 of the IBC for elevator evacuation and Section 3001.3 for reference to ASME A17.7).

Another aspect that has evolved within the codes is the concept of a Fire Service Access Elevator (FSAE). The IBC and ASME A17.1 (elevator code) currently require certain features for use by the fire service. Additionally, the IBC has more rigorous requirements for elevators for FSAE that involve a lobby with upgraded construction requirements, stairs adjacent to or a protected path to such lobbies and standpipe locations and other related requirements. This particular package of requirements is only required for buildings with occupied floors greater than 120 feet (36.6 m) above the lowest level of fire department access (see Section 3007 of the IBC).

This topic is specifically addressed in the SFPE/ICC *Engineering Guide—Fire Safety for Very Tall Buildings* (2021).

Acceptable methods

Means of egress. Chapters 10 of the IBC and IFC are acceptable methods, given that they are based on the factors used in the development of the performance code.

Accessibility. Chapter 11 of the IBC and ICC A117.1 would be considered the acceptable methods. Because of the controversial nature of these provisions, a prescriptive approach should be undertaken as a minimum. At the very least, the objectives, functional statements and performance requirements were drafted to capture the intent and purpose of accessibility provisions.

Transportation systems. The acceptable methods for these provisions include Chapter 30 of the IBC, ASME A17.1/B44, A17.7-2007/CSA B44.7, ASME A90.1 and ASME B20.1. Note that ASME A17.7-2007/CSA B44.7, *Performance-Based Safety Code Elevators and Escalators*, is actually a performance elevator standard and more directly fits within the structure of this code but, as discussed in

Section 103, standards are typically not referenced in this code. This standard takes high-level performance objectives and assists in creating practical solutions. As with this code the document allows compliance with prescriptive methods as well. In fact, A17.1/CSA B44 references the standard directly. The standard has a third-party process built into the document.

CHAPTER

7 SAFETY OF USERS

SECTION 701—HAZARDS FROM BUILDING MATERIALS

This section is concerned with injuries to building occupants related to the building materials themselves. There are two areas this section addresses:

- Contact with glass (glazing).
- Emission of liquids, gases, radiation or solid particles from building materials.

Glass is used in many occupancies because of its architectural appeal. Depending on how glass is used within a building, it can be hazardous to occupants during everyday use of the building. Glass walls may be mistaken for a path of travel, or the glass can be subject to breakage due to its location and arrangement. Other hazards related to building materials include harmful fumes from carpeting or the use of materials such as asbestos. Typically, the Consumer Products Safety Commission (CPSC) or federal or state level Environmental Protection Agency (EPA) rules regulate such hazards.

Section 701.1 Objective

The objective of this provision states that the code intends to protect occupants from injury resulting from the impact of glass or other transparent materials or injury resulting from the breakage of such materials. Additionally, as noted above, this section may also be applicable to such hazards as asbestos in buildings and off-gassing from carpet materials. The use of such materials tends to be more of an issue with existing buildings, as many of these hazardous building materials are not available on the market for construction of new buildings.

701.2 Functional statement

To protect the occupants, glass and similar materials must be used in a manner that avoids the risk of injury. Such injury can come from impact itself or from the breakage of such material, resulting in sharp edges likely to cause cuts. Also, with regard to the other building materials hazards such as asbestos or perhaps insulation, the approach could be to prohibit or perhaps properly seal off the leakage of dangerous chemicals or release of particles.

701.3 Performance requirements

The use of these materials is outlined in the performance requirements. Glass or any other brittle material that building occupants may come in contact with shall break upon impact in a manner unlikely to cause injury. The typical approach is to use tempered glass or other equivalent material able to resist impact. Protection from impact may be via guards or other types of barriers. Avoiding harmful concentrations of materials may necessitate a proper seal, proper ventilation or perhaps avoidance of material use.

SECTION 702—PREVENTION OF FALLS

This section prevents people from falling over the edge of a floor surface where there is an abrupt change in floor elevation and prevents people from falling from openings in a building's envelope (such as window openings in a wall or a roof surface). Because the prescriptive codes have usually required this edge protection where there has been a change in elevation of 30 inches (762 mm) or more, that threshold has been carried over to this performance code. Essentially, in this case, there is only one design performance level provided.

702.1 Objective

The threshold used in the prescriptive *International Building Code®* (IBC®) is 30 inches (762 mm). Therefore, the objective relates to the need to protect people from the risk of falling more than 30 inches (762 mm). This means that the performance code, like the prescriptive code, does not intend, generally, to prevent falls on surfaces less than 30 inches (762 mm) apart vertically. However, Section 601.3.5 requires that the means of egress take into account human biomechanics and expectation of consistency. Therefore, the performance re-quirements relate to the performance of the guard, not the building pedestrian system.

702.2 Functional statement

The key to this functional statement is "unintentional." It would require sophisticated and expensive technology to prevent someone from falling who is determined to do so and would be outside the scope of current prescriptive codes.

702.3 Performance requirements

The performance requirements are such that some sort of protection must be provided that prevents people from going over an edge of a floor surface that is more than 30 inches (762 mm) higher than an adjacent floor surface. The protection must be sufficient to protect against falls by small children as well as adults. Therefore, the protection must be able to withstand the force of an adult as well as withstand the attempts by a small child to somehow squirm through the protection. The concern of children passing through is addressed in the prescriptive code by limiting openings such that a 4-inch (102 mm) diameter sphere cannot pass through any openings to a height of 34 inches (864 mm). The sphere represents a small child's head, and the height reflects the height of the children that are being protected.

This section also addresses roofs that have permanent access, such as a hotel roof that contains a pool. These roofs are also termed "occupiable roofs" by the IBC.

SECTION 703—CONSTRUCTION AND DEMOLITION HAZARDS

This section requires that the public, personnel on a construction site and property adjacent to a construction site be protected from the hazards typically associated with demolition and construction operations. The provisions parallel those of Chapter 33 of the IBC, the *International Fire Code®* (IFC®) and Chapter 15 of the *International Existing Building Code®* (IEBC®).

703.1 Objective

The objective statement in this section is self-explanatory. It parallels the intent of Chapter 33 of the IBC, which is to require protection of people and property from the various hazards imposed by construction or demolition operations.

703.2 Functional statement

Functional statements give the user specific information regarding what could possibly be of concern regarding a construction or demolition operation. Namely, excavation operations and large lift operations can damage adjacent property. Nearby pedestrians can be injured without proper precautions. In addition, the provisions address the "attractive nuisance" hazard associated with construction sites by requiring some level of protection against the entry of unauthorized individuals. Finally, prevention measures to avoid hazards such as fires or explosions are necessary. Fires are more likely to occur during construction than during normal use of the building. See the User's Guide discussion for Section 703.3.

703.3 Performance requirements

The hazards associated with construction or demolition operations are related to lift operations, stability of scaffolding, movement of personnel, excavation, other operations that impact an adjacent property and the ability of a partially completed structure to resist natural hazards such as wind and rain.

Unique fire hazards can exist during construction and demolition operations. Most are associated with the temporary storage and use of flammable and combustible materials on the job site. Even in buildings of noncombustible construction, it is common to find stores of lumber for forms or bracing that represent a significant fuel load. Many finishes and adhesives are flammable in an uncured state and represent a fire hazard if not used in accordance with good safety practices. Also, during construction or demolition, the normal fire safety features of the building are not functional. Fire-resistant construction features are incomplete, and detection or suppression systems are often not operational. Temporary equipment such as space heaters, work lights or welding equipment, and cooking on construction sites represent sources of ignition and should be used carefully.

SECTION 704—SIGNS

This section provides criteria for the incorporation of signage into a building or facility. In the prescriptive code, signs are addressed by individual sections by providing provisions for specific features of a building or structure. An incidental description of the physical characteristics and desired placement of the sign often accomplish this. The performance requirements for signs are centralized and apply to any building feature requiring signage, whether that be an exit sign or a marking of a specific feature such as the storage of hazardous materials or an accessible route.

SECTION 705—EMERGENCY NOTIFICATION

This section addresses the need for the notification for some manual action to preserve the safety of people or limit damage to a building or structure or its contents. Some of the conditions that might warrant emergency notification include fires, storms (tornado, hurricane, severe thunderstorm), bomb threats, hazardous materials releases, hazardous conditions from utilities and any other condition that could lead to injury or property damage.

705.1 Objective

The objective statement identifies the requirement that emergency notification systems be provided to initiate manual intervention needed to limit hazards to people or property.

705.2 Functional statements

The first functional statement introduces the concept that notification should occur in a timely manner so that manual action can be taken without harming those taking the action.

The second functional statement discusses the need to provide sufficient information to emergency responders so that they can identify, locate, and mitigate hazards efficiently and safely.

705.3 Performance requirements

The means of notification need to be effective for everyone being notified. Thus, if people with hearing impairments are among those to be notified or if high ambient sound levels are expected, the means of notification must not depend solely on audible information. If the egress plan involves phased evacuation or relocation to safe places within the building, means must be provided to maintain communication to all those who may have to be evacuated later.

In some occupancies, people are expected to be sleeping at times, and notification systems must be designed so that there is reasonable assurance that these people will be awakened. Notification might not be an alarm system per se. For example, notification might be via the odor built into liquefied petroleum gas. Additionally, detection and notification of a failure could be through the observation and reporting of staff. Not all buildings warrant the use of notification devices.

Acceptable methods

Hazard from building materials. Following Chapter 24 of the IBC is one way of achieving the objectives, functional statements and performance requirements of this section with regard to glass. Note that the IEBC does prohibit the use of materials such as asbestos in Chapter 3 of that code. Materials that have been banned by the CPSC or EPA and are no longer on the market are more likely to be found in existing buildings.

Prevention of falls. Refer to Chapter 10 of the IBC.

Construction and demolition hazards. Refer to Chapter 33 of the IBC, Chapter 33 of the IFC and Chapter 15 of the IEBC.

Signs. The acceptable methods for this particular section are included throughout the building and fire codes. More specifically, Chapters 10 and 11 of the IBC have provisions directly related to signage. The IFC has provisions throughout related to signage.

Emergency notification. First, the fire and building codes do not require notification in all cases. The need for notification in a performance-based design will be based on whether the occupants need to be notified of a problem in order to provide them with enough time to take specific actions. The IBC and IFC have requirements for a mass notification risk assessment for certain occupancies including college campuses and primary and secondary education facilities. These assessments will result in different notification needs.

Various standards provide guidance on the number and distribution of audible and visible devices that are effective at notifying people in emergencies. Information on the conveyance of meaningful messages is less straightforward. If common devices are used for a range of emergency conditions, some means of differentiating the hazard is needed if the action to be taken is different. The pulsing of notification devices in a specific pattern has been used, but this requires training and signage to be effective. In recent years, new methods of transmitting textual information have been developed, but these must also account for the fact that some people may not understand English instructions. All these possibilities should be considered in the design of an effective notification system.

Refer to Chapter 9 of the IBC and IFC for fire alarm requirements and Chapter 4 of the IBC for emergency alarms.

CHAPTER

8 MOISTURE

SECTION 801—SURFACE WATER

This section presents requirements to prevent local drainage and storm-water runoff from entering buildings or moving across property lines and causing damage to properties. Because water may become contaminated, streams and other bodies of water must be protected from water suspected of being exposed to hazardous materials stored at building sites.

The prescriptive code does not contain a section dealing with water runoff from one property to another. However, many local ordinances contain sections dealing with this problem. Water leaving property comes not only from rain but also from other sources such as irrigation or firefighting. Water used in firefighting can mix with hazardous materials, causing pollution of streams and lakes. Sections in the *International Building Code*® (IBC®) and *International Fire Code*® (IFC®) address this concern along with the secondary containment requirements for hazardous materials.

Another problem arising from uncontrolled water runoff is property damage and nuisance. Unless water is controlled when leaving a property, serious damage can result to property on the outfall side. Proper grading can alleviate this problem.

Surface drainage must be constructed to prevent the formation of a blockage, and in the event of a blockage, the obstruction must be able to be easily removed.

801.1 Objective

The object of this section is to protect people from injury or illness who have been affected by improper surface water damage from pollution. It is also meant to protect property from damage occurring from accumulated surface water.

801.2 Functional statements

Surface water damage is prevented through approved methods of construction and site layout.

801.3 Performance requirements

Compliance with the performance requirements of this section is the responsibility of the design engineer. It is the engineer's responsibility to design the site grading so storm water and local drainage are safely conveyed from the site without causing damage to adjacent properties or erosion of drainage paths.

SECTION 802—EXTERNAL MOISTURE

The intent of this section is to prevent moisture originating at the exterior of the building from adversely affecting the occupants' health and safety and the structural and functional performance of the building. This section deals with both the liquid and vapor forms of water. If liquid penetrates through condensation or as a liquid, such moisture may cause the decay and corrosion of building elements. Also, the damp climates created by the penetration of moisture into the building may promote bacteria growth and be harmful to the occupants. The main mechanism to achieve this goal is the prevention of water and water vapor from entering the building envelope. However, should water vapor penetrate the exterior skin, it may be dealt with in several ways to achieve the intended performance. For example, a hygroscopic insulation may be installed within the exterior wall cavity such that the condensed vapors are sufficiently absorbed and do not adversely affect the building components. Or, the wall may be ventilated to remove the vapor before it can cause damage.

The prescriptive code shows moisture provisions in the IBC Chapter 14, Exterior Walls; Chapter 15, Roof Assemblies and Rooftop Structures; and Section 1805, Dampproofing and Waterproofing. The concepts are much the same as the current prescriptive code, though the aspect of occupant protection is probably more apparent in the performance provisions. The prescriptive code appears to address only property.

There was some discussion that this particular section should be combined with the surface-water provisions as, generally, the intent is to avoid water contact with and penetration into the interior of the building. However, there is a significant difference in how water vapor is handled to avoid damage or mold.

802.1 Objective

This objective specifically protects the building occupants from the effects of moisture penetrating the building envelope. This penetration can be leaks that cause slip hazards, or moisture promoting the growth of mold and mildew within the building. An additional objective is to protect property from damage. More specifically, the concern is with water reaching building elements and material being destroyed through direct contact with liquid or vapor. In the case of wood, contact with moisture can result in rotting and possible structural failure. Property protection also extends to nonstructural elements.

802.2 Functional statement

The functional statement dictates that to achieve the objective, the building needs to provide a specific function. It must be constructed in a manner that will prevent penetration into and accumulation of water or water vapor in or on the building. Penetration can be avoided by the use of techniques such as weather-resistant barriers and flashing, or vapor retarders. The function can simply be to reduce the amount of water remaining in contact with the building surface. This may include good drainage through sufficient slope of the roof and appropriate soils located underneath a structure to remove moisture.

802.3 Performance requirements

There are several performance requirements for this section to cover all areas of possible moisture contact with the building envelope. Basically, these provisions break down the functional statement into more specific components.

802.3.1 Water penetration

The roof and exterior walls in contact with moisture are protected from initial precipitation and from vapor with moisture barriers and flashing. Water accumulation on roofs is avoided through proper slope and/or drainage (IBC Chapters 14 and 15).

802.3.2 Building elements in contact with the ground

The protection of elements in contact with the ground, including walls, floors and structural elements, must be provided. Depending on the water level, these components are to be waterproofed or dampproofed as required in the current codes (IBC Section 1805).

802.3.3 Concealed spaces and cavities

Moisture within a concealed space must be contained so as to prohibit contact with structural elements. Such contact can be through liquid or vapor. This may be caused by an inappropriately sealed soffit (IBC Chapters 14 and 15).

802.3.4 Moisture during construction

The construction process itself should not allow moisture from the building materials or the atmosphere to cause any permanent damage to structural or nonstructural building elements.

SECTION 803—INTERNAL MOISTURE

Generally, this section of the code is concerned with water sources within the building having a negative effect on the safety of the occupants or the structural stability of the building and causing general property loss within the interior of the building. Again, as with external moisture, this includes both liquid and vapor forms of water.

In terms of occupant safety, the concern is with vapor buildup as a result of improper ventilation and with surfaces such as showers where fungal growth can affect the health of the occupants. Surfaces that come into contact with water must not promote the growth of mold or mildew. This requires such surfaces to be impervious and easily cleanable. Property loss occurs, for example, when a sink overflows and water damages structural and nonstructural elements of a building. Also, such overflow can lead to the growth of mold and mildew because of contact with inappropriate surfaces.

The current codes cover such issues within the *International Plumbing Code*® (IPC®) Chapter 4, Fixtures, Faucets and Fixture Fittings; the IBC Chapter 12, Interior Environment; and the *International Mechanical Code*® (IMC®) Chapter 4, Ventilation.

Chapter 4 of the IPC provides the requirements for allowed plumbing fixtures and how they are to be installed. This includes specifying elements such as impervious and easily cleanable surfaces. In addition, the IPC requires walls above built-in tubs with installed showerheads to be constructed of smooth, noncorrosive, nonabsorbent and waterproof materials.

In addition to the type of surfaces, IPC Chapter 4 has requirements for drain sizes to avoid overflow of water beyond the plumbing fixture, which can lead to damage or unwanted moisture.

The use of ventilation is another way to combat internal moisture. Chapter 12 of the IBC requires ventilation in accordance with the IMC. Chapter 4 of the IMC sets specific exhaust rates for particular use groups and areas within those use groups. For example, for toilet rooms and bathrooms within one- and two-family dwellings, 50 cfm (923 L/s) intermittent or 20 cfm (9.4 L/s) continuous airflow is required. This serves to remove odors and moisture from the enclosure.

803.1 Objective

The objective reflects the overall intent of the provisions, which is to protect the occupants from the negative effects of internal moisture related to the buildup of bacteria and to protect against potential property damage caused by water overflow and splashing.

803.2 Functional statement

The functions of the building must include the avoidance of moisture accumulation on surfaces, which can lead to contaminants and fungus. If such accumulation cannot be avoided, then the surfaces must not promote such growth. In addition, prolonged exposure to a moist atmosphere can affect the integrity of both structural and nonstructural elements. Water that overflows must not cause property damage to the neighboring occupancy. Therefore the building, or more specifically the fixtures and surrounding areas, must be able to contain the water in an appropriate manner. Another aspect covered by internal moisture is splashing. Any area surrounding a plumbing fixture where water may come in contact with a surface on a regular basis is susceptible to the growth of mold and mildew, and thus the surface must be appropriate.

803.3 Performance requirements.

To perform the necessary functions called for in the functional statement, several performance requirements have been set forward.

803.3.1 Excess moisture removal and protection

This section requires both thermal resistance and ventilation to control moisture in any space within the building. The level provided depends on the use of the space. This is done by the prescriptive code through the breakdown of ventilation flows per the use of the space. Each space will need evaluation because prolonged exposure to moisture can cause permanent damage. For example, wood tends to rot if exposed to excessive moisture.

803.3.2 Overflow

This requirement pertains to overflow from fixtures such as sinks or bathtubs flowing into areas where damage may occur to adjacent occupancies. The strategy is to avoid an overflow rather than manage the effects of an overflow. Avoiding overflow involves utilizing a particular size drain and an overflow drain.

803.3.3 Floor surfaces

Floor surfaces in areas where sanitary fixtures or laundry facilities are located should be impervious and easily cleaned. This requirement protects occupants from the growth of fungus or the accumulation of contaminants in areas subject to water overflow or to splashing resulting from use. In addition, such growth can lead to foul odors.

803.3.4 Wall surfaces

As with floor surfaces, wall areas in contact with water from sanitary fixtures and laundry facilities must be made of impervious materials and linings and be easily cleanable. The same reasons as noted for floors apply to this section.

803.3.5 Surfaces and building elements

This requirement emphasizes the negative effect splashing fixtures may have. It includes walls, floors and any other elements of a building likely to come in direct contact with splashing. An example beyond walls and floors is a cabinet adjacent to a sink.

803.3.6 Water splash

In addition to having surfaces that resist the effects of splashing, areas must be designed to prevent water from penetrating behind the lining or entering into other confined spaces.

Acceptable methods

Surface water. Acceptable methods to prevent problems from surface-water runoff fall mainly into the realm of good engineering practice. In terms of contaminated runoff, the prescriptive fire, building and plumbing codes address this issue indirectly through requirements for spill control and secondary containment and requirements for release of waste.

Exterior moisture. As noted, the acceptable methods for protection against external moisture are found in IBC Chapters 14 and 15, and Section 1805. These provisions offer prescriptive methods that encompass the entire outer building envelope. IBC Chapter 14 provides measures to prevent moisture from penetrating the walls; IBC Chapter 15 provides measures to prevent moisture from penetrating the roof; and IBC Chapter 18 provides protection from water originating beneath the structure.

One issue not addressed in the current prescriptive documents is the construction process, specifically, moisture content of the construction elements.

Interior moisture. As noted, the main methods of the prescriptive codes include Chapter 4 of the IPC, Chapter 12 of the IBC and Chapter 4 of the IMC.

Chapter 4 of the IPC specifically provides nationally recognized standards for plumbing fixtures such as sinks and bathtubs.

CHAPTER 9 INTERIOR ENVIRONMENT

This chapter describes the performance provisions of a building's environment. It is divided into four areas:

- Climate and Building Functionality addresses the provisions of space and climate appropriate for the needs of the occupants, the activities and the furnishings.
- Indoor Air Quality addresses the need to provide adequate clean air within a building or facility for the occupants. This includes controlling the moisture content, odors, poisonous fumes and any other air quality issues. This section is aimed at such problems as "sick building syndrome."
- Airborne and Impact Sound requires that tenant, common and habitable spaces be insulated against sound transmission.
- The criteria for light for everyday use in habitable spaces and means of egress is provided in the Artificial and Natural Light section.

Acceptable methods

This chapter represents the comfort and health issues addressed by the prescriptive *International Building Code*® (IBC®) in Chapter 12. The artificial and natural light requirements relate in part to portions of Chapter 10 of the IBC. The *International Mechanical Code*® (IMC®) also supports appropriate ventilation through a variety of applications pertaining to indoor air quality. Minimum ventilation rates are addressed in Chapter 4 of the IMC. Related standards include ASHRAE 62.1, *Ventilation and Acceptable Indoor Air Quality*, and 62.2, *Ventilation and Acceptable Indoor Air Quality in Residential Buildings*.

CHAPTER

10 MECHANICAL

This chapter provides the performance provisions for the building mechanical system(s). It addresses the installation and functions of HVAC, refrigeration and piped services. Though indoor air quality is supported by minimum ventilation rates, the provisions of this chapter focus on the performance of the equipment rather than indoor air quality. Chapter 9 provides more guidance on indoor air quality and climate.

The ventilation section contains performance requirements for habitable and concealed spaces. The need for combustion air and disposal of contaminated air are also covered. These issues are prescriptive *International Mechanical Code*® (IMC®) issues that interface with the prescriptive *International Building Code*® (IBC®). It was believed that placing these items within a single group title would better demonstrate how the current codes are formatted.

Acceptable methods

The IMC is an acceptable prescriptive method related to the objectives of this chapter.

CHAPTER 11 PLUMBING

Chapter 11 contains performance provisions that relate to the prescriptive plumbing requirements of both the *International Plumbing Code*® (IPC®) requirements and, to a certain extent, the *International Residential Code*® (IRC®) for single-family dwellings. The provisions cover personal hygiene, laundering, domestic water supplies, wastewater and nonpotable water system. Accessibility must be addressed for these provisions based on the application of Section 602.

The provisions for personal hygiene within Section 1101 require that proper sanitary fixtures, and such things as adequate toilet facilities, be provided for the occupants of a building to maintain health and to prevent the spread of disease.

Section 1102 requires adequate laundry facilities for occupant use. The number of fixtures required may vary, for example, based on whether the building is in an apartment or single-family home. The requirement is merely that proper facilities be required. This may be a central laundry facility in an apartment building, whereas every single-family home would require the hook-ups and space necessary to accommodate a washer and dryer. Chapter 10 addresses dryer installations.

Generally, Section 1103 deals with sanitary distribution of water for drinking, food preparation and hygiene. More specifically, Section 1103.3.8 requires the installation of backflow prevention valves to safeguard the water supply.

Wastewater is addressed by Section 1104, which provides for a safe and properly functioning sewerage system within a building or group of buildings. It requires that sewer gases be prevented from entering the building. Section 1104 also addresses requirements such as proper installation of piping, including provisions to unstop the plumbing system.

Section 1105 facilitates the safe and proper use of nonpotable water systems. This allows a more sustainable use of water resources that has been recognized by the IPC since 2012.

Acceptable methods

The IPC is an acceptable prescriptive method related to the objectives of Chapter 11 of this document. In the case of single-family homes, the plumbing provisions of the IRC would apply. Note that Chapter 29 of the *International Building Code*® (IBC®) duplicates the fixture count requirements of the IPC.

CHAPTER

12 FUEL GAS

SECTION 1201—FUEL GAS PIPING AND VENTS

1201.1 Objective

This section ensures that fuel gas is used in a manner that does not create a hazard to the occupants. This primarily means that the fuel gas be used in a manner that does not permit unsafe levels of combustible byproducts or make it a source of ignition of an unwanted fire.

1201.2 Functional statement

When fuel gas is used as an energy source, the installation must be safe. This applies to all aspects of its use, including installation of an appliance, piping system, venting, oxygen depletion safety shutoff system (ODS), safety controls, clearances from combustibles, etc.

1201.3 Performance requirements

Many issues must be addressed in order to ensure that fuel gas is used safely as an energy source. Safe operation of appliances requires that supply systems be properly sized to deliver fuel gas at the pressure required by the appliance. An automatic shutoff system must be installed to safeguard occupants from unsafe conditions should the appliance malfunction.

Gas appliances require adequate combustion air for proper operation when multiple gas appliances are installed in the same space, and care must be taken to ensure there is enough combustion air for all appliances operating simultaneously. Gas appliances may be vented or unvented depending on the appliance's listing. To remove the products of combustion, appliances will be vented to the exterior of the building by the use of flues. When multiple gas appliances are connected to the same flue it must be done so that there is no spillage of exhaust gas. The products of combustion for unvented appliances will not be vented to the exterior of the building by the use of flues but will be vented into the atmosphere of the room or space. Gas appliances must be provided with individual shutoff valves so that they can be safely maintained, tested or repaired without shutting off the main supply. Gas appliances must be tested and maintained according to the manufacturers' instructions. The supply system in the building must be designed and installed to prevent adverse effects on the utility's supply or other users connected to the system.

Acceptable methods

The *International Fuel Gas Code*® (IFGC®) is an acceptable prescriptive method related to the objectives of this chapter.

CHAPTER

13 ELECTRICITY

SECTION 1301—ELECTRICITY

1301.1 Objective

The objective of this section is to provide for the installation of electrical services and equipment in a manner that minimizes the risk of shock or electrocution to people and minimizes the possibility that such systems or equipment will start a fire.

1301.2 Functional statement

The installation of electrical services and equipment should follow strict procedures in order for the services to operate safely. These procedures were developed to ensure that safeguards against shock, electrocution and fire hazards are in place.

1301.3 Performance requirements

Shock hazards are frequently associated with accidental contact with energized parts of electrical equipment, or situations in which such parts come in contact with conductive building elements, which then become electrically charged. Live parts of electrical services and equipment are generally isolated from accidental contact by being enclosed in insulation or grounding conductive materials.

Fire hazards usually involve faults or failures that result in excessive current creating heat or sparks that ignite nearby combustibles. Protection against such faults is usually provided by properly rated overcurrent protective devices on distribution circuits and in equipment. Protection against sparks is provided by enclosures.

Electrical equipment generally produces heat as a byproduct of normal operation. Such equipment is provided with installation and safe operating instructions that communicate the need for ventilation so as to prevent excessive heat buildup that can lead to fire. Also, using appliances that require a higher current than is available will result in resistance heating, which can be a potential source of ignition.

Electrical equipment installed in locations containing flammable or explosive materials requires special design and certification to prevent fires or explosions. Generally, the prescriptive code, NEC, would require specifically classified electrical equipment within certain areas.

The operation of some electrical services and equipment are themselves directly related to life-safety. Examples include systems that provide life support to patients in hospitals or power to fire-safety systems. Such systems often require continuous or essential power systems that provide the needed reliability in the event of an interruption of the primary system. Note that this chapter is focused on safety when using electrical systems. The need for continuous (standby or emergency) or essential power will be triggered by other aspects of this code.

Electrical installations must follow practices that maintain the integrity of grounding systems and overcurrent protection that begin with the systems provided by the utility.

Acceptable methods

With adherence to the provisions of recognized electrical codes such as NFPA 70: *National Electrical Code* and industry practices, work performed by licensed practitioners, permitting and inspections, and adherence to installation and use instructions are usually sufficient to ensure electrical safety.

CHAPTER

14 ENERGY EFFICIENCY

SECTION 1401—ENERGY EFFICIENCY

1401.1 Objective

The energy efficiency chapter provides the requirements for building systems and portions of a building that impact energy use in new construction, and it promotes the cost-effective use of energy. This section permits flexibility in the application of code provisions as long as performance requirements are met. Note that this chapter is based on the 2021 and earlier editions of the *International Energy Conservation Code®* (IECC®) which had not yet addressed energy generation and sustainable energy technologies.

1401.2 Functional statement

Energy efficiency is needed to conserve depletable energy sources. The cost of the required action compared to the energy that will be saved over the life of the action must be considered when applying conservation methods. Therefore, the building must be designed to take these concerns into account.

1401.3 Performance requirements

1401.3.1 Energy performance indices

To provide for the efficient use of depletable energy sources, the building envelope must be designed and constructed within stated parameters called "energy performance indices." These indices indicate the amount of energy from a depletable energy source needed to maintain the building at a constant internal temperature, measured per square foot (m^2) and per degree-day under standard conditions. These indices are based on the region of the country as well as the use of the building. In some cases, the local jurisdiction may choose not to specify energy performance indices for certain types of buildings.

1401.3.2 Temperature control

For buildings requiring a controlled temperature, the building design and construction must take into account various factors. Normally, only insulation, types of windows, etc., are considered when addressing energy conservation. However, to provide for the efficient use of energy, there are several other items that need to be taken into consideration such as thermal resistance, solar radiation, air tightness and heat gain or loss from building services.

Acceptable methods

The IECC is an acceptable method related to the objectives of this chapter. The IECC provides both prescriptive and performance-based approaches to achieving energy efficiency.

CHAPTER 15 FIRE PREVENTION AND IMPACT MANAGEMENT

There are two ways to deal with fire: it can either be prevented or managed.

SECTION 1501—FIRE PREVENTION

Prevention is an essential emphasis of fire codes. It is also recognized that the expectation is to limit unwanted ignition to an acceptable level. It is unreasonable to believe that all unwanted ignition can be eliminated.

Section 1501 addresses both temporary ignition sources, such as welding, and more permanent installations, such as water heaters. Section 1501 focuses on building equipment and systems and fuel load interaction with ignitions sources. The inclusion of building equipment and systems requirements were influenced by codes such as the *National Electrical Code*, the *International Fuel Gas Code*® (IFGC®) and the *International Mechanical Code*® (IMC®). The provisions of Section 1501 addressing fuel load interactions with ignition sources are focused more on the fire code provisions, which typically place greater emphasis on preventing ignition than does the building code.

This section of the performance code is strongly linked to Chapter 16, Management of People, as it recognizes that many fires can be prevented by adequate training and safety procedures. Likewise, public education has a significant impact on individual awareness of fire issues, hence the likelihood of people exhibiting fire-safe behavior.

The purpose of this section is to reduce the potential of permanently installed building equipment, appliances and services to cause a fire because of their installation. All types of permanently installed equipment, appliances and services represent potential ignition hazards and need to be installed in a fashion that minimizes or prevents these hazards from occurring. Fuel-burning equipment, including gas, oil and solid fuel-fired types, can, by their nature, transfer heat to building materials when located too close to such materials. Electrical equipment produces heat and potential sparks, thus requiring clearances to combustible materials and hazards such as areas where flammable vapors are likely. Additionally, should a spark or flame escape the equipment or appliance enclosure, ignition of building materials or contents can occur. These provisions were developed with the potential hazards of this equipment in mind.

Because many fires can be prevented by exercising appropriate fire-safe behaviors or practices, it is essential that procedures and training for occupants to facilitate fire-prevention objectives be developed. These must be documented appropriately where they are integral to the strategy for meeting the objective and should include appropriate expectations for follow-up and monitoring.

1501.1 Objective

This section contains provisions related to several aspects of preventing ignition. These provisions intend to prevent the ignition of building materials caused by permanently installed equipment, appliances or services and from portable equipment or appliances, which are connected by cord and plug. More permanently installed equipment is typically regulated by construction codes and more temporary appliances and equipment are typically regulated by the codes such as the *International Fire Code*® (IFC®). In addition, the requirements are also focused on hazards such as welding and cutting operations, hazardous materials processes and smoking. The prescriptive regulatory solutions will likely be found in several locations due to the scoping of such issues. Construction codes such as the *International Building Code*® (IBC®), *International Existing Building Code*® (IEBC®), IMC, IFGC and the NEC will regulate permanent installations of equipment and systems. Codes such as the IFC, which contains many operational requirements, will address portable appliances and equipment and also hazardous materials and operations.

1501.2 Functional statement

The main functional statement generally encompasses all ignition hazards, but the two more specific functional statements address the need for fuel-fired or electrical equipment, appliances or services to be installed in a manner that reduces the potential for the installation to be a source of ignition. Fuel-burning appliances and electrical equipment are generally intended for installation in accordance with the manufacturer's instructions. These instructions should be provided as part of the design submittal to determine if the equipment is being installed in accordance with the manufacturer's recommendations. If the equipment has not been listed or independently evaluated for safety, it is up to the designer to demonstrate to the code official that the equipment installation provides an acceptable level of safety.

To avoid the equipment becoming a potential source of ignition, several factors need to be considered. Appliances must not generate temperatures or operate under conditions that could ignite nearby combustible materials. Electrical conductors and components must not create a condition in which they will overheat and ignite combustibles either in or near the appliance, or in the branch circuit wiring supplying the product. Additionally, the equipment must provide an acceptable level of performance under normal operating conditions and any other conditions that may be encountered during the life of the product.

Standards for safety have been developed for most fuel-burning and electrical appliances and equipment by organizations such as UL and ASTM. These standards provide a comprehensive set of safety criteria with which to evaluate the ability of the equipment to not produce an undesired source of fire ignition. These standards are customized to address safety conditions that are unique to the specific appliance or equipment.

Equipment that has been listed by an approved testing and certification laboratory has undergone an independent safety investigation. As part of the listing, the certification agency evaluates the equipment and certifies that it complies with appropriate safety standards when installed in accordance with the manufacturer's installation instructions. Equipment installed within the limitations of its listings and in accordance with the manufacturer's installation instructions can generally be considered to not serve as a source of fire ignition.

It is also necessary to periodically test and maintain appliances and equipment in accordance with the manufacturer's instructions and in accordance with any referenced installation and use standards to ensure that the appliances and equipment do not create a potential source of fire ignition during the life of the product.

When it comes to less permanent and general ignition hazards, control of heat/energy sources to limit the occurrence of unwanted ignition would be accomplished, for example, by controlling open burning, open flames, smoking, torches for removing paint, electrical safety (extension cords), asphalt kettles and unsafe chimneys. Suitable controls and procedures must be identified in meeting the objectives of other sections of this document to prevent ignition. Providing and using equipment that is suitable to the environment they will be exposed to, and in compliance with the listing, are basic expectations to meet this objective. Similarly, maintaining all equipment so that a fire hazard does not exist is important.

1501.3 Performance requirements

The performance requirements are intended to provide more detailed intent of the provision. These provisions will likely link more closely with the prescriptive codes or approved documents. The first performance requirement is focused on built-in equipment and systems integral to the building.

The other two focus on reduction of fuel sources and the interaction of fuel and ignition sources

It is the designer's responsibility to assess the ignition sources and take whatever actions are necessary to develop a solution that will limit these sources. Though the goal should be to eliminate these sources, it is recognized that it may not be possible to completely eliminate all sources in a cost-effective manner. Therefore, a balance of preventing ignition and managing fire is necessary.

In all cases, the fire risk or probability of ignition, as well as the impact of even a minor fire, must be considered. Special consideration needs to be given to buildings or facilities based on their respective performance groups as defined in Chapter 3.

In addition to the ignition source being controlled, the fuel hazards can be controlled by limiting the quantity, composition or configuration of the material so that the occurrence of unwanted ignition is limited. This can be accomplished, for example, by specifying requirements for interior finish, furnishings, cleaning of commercial cooking exhausts, accumulation and storage of combustible waste and other combustible fuel loadings (Christmas trees). These examples are not intended to be exhaustive.

It is recognized that in some cases both ignition sources and fuel hazards may be present. An appropriate strategy to meet the objective may be to provide and maintain barriers or separations between the heat ignition source and the fuel supply, thereby limiting the effects of radiant, convection and conduction heating.

Examples of employing the strategy of maintaining barriers or separations would be the prescriptive provisions for fire-resistant-rated assemblies, electrical panel clearance requirements and electrical wiring clearance requirements. Again, these examples are not intended to be exhaustive. This strategy must also be extended to things such as controlling the access of unauthorized individuals to the location of fuel hazards. This can be accomplished, for example, by securing vacant buildings or controlling access to certain areas of the building or facility. Where security cannot be guaranteed in instances such as these, then all fuel hazards should be removed. Consideration should also be given to limiting untrained individuals to certain activities or locations. In addition, the maintenance or respective safeguards should take into account accidental occurrences such as vehicle accidents and impacts that can render well-designed separation strategies inoperative.

The heightened security that has been added in and around high-rises and stadia since the September 11, 2001, World Trade Center disaster is another example of a fire prevention solution to factor into a performance design. Indeed, even the security at airports can be considered part of a system of fire prevention protection for our buildings and infrastructure. As discussed in Section 305, building and fire codes have not usually addressed issues such as terrorism. Therefore, fire prevention related to such anticipated ignition sources will need to be explicitly acknowledged as potential events to explain the various fire prevention measures taken.

SECTION 1502—FIRE IMPACT MANAGEMENT

General

Section 1502 provides objectives, functional statements and performance requirements for managing the impact of a potential fire event that can occur in a building or facility. This section assumes that a fire can occur in the buildings or facilities despite any "prevent fire" measures that are being taken. The objective is to limit the impact of a fire to an acceptable level on the occupants, the general public and the building or facility, including its contents, use and processes. Other inherent objectives in this section are to provide some level of protection for the building or facility's mission and revenue stream and the community tax base. In general, those are the underlying themes of building and fire regulations in the United States. Often, insurance companies may require higher performance due to the cost associated with insuring the building or facility or as a way to reduce premiums. In some countries the focus of the regulations is more about life safety and more emphasis is placed on the owner and insurer for property protection. Generally, the prescriptive codes have been seen as minimum requirements.

1502.1 Objective

This section is intended to reduce the likelihood of death or injury to an acceptable level to those persons involved in a fire within the building or facility. It also prevents or reduces damage or loss of property because of the spread of fire within a structure or to an adjacent structure. "Persons involved" include building occupants, emergency responders and people in the vicinity of the building or facility. Consideration is also given to the impact a fire may have on the use of the building or facility, including any process associated with the building or facility. The objective statement is a continuation of the basic principle of building and fire codes from the beginning of code development. The difference is that this format provides the opportunity for the reader to understand the overall intended performance of a particular building or facility in a fire event and how different subsystems interact to achieve the desired objective.

1502.2 Functional statements

The general functional statement sets a single performance level for fire life safety but sets multiple upper limits for property damage based on the performance group classification of the facility.

A functional statement of particular interest is Section 1502.2.8, which places specific emphasis on providing a level of fire safety for occupants with physical or mental disabilities that is comparable to that provided for those who do not have such disabilities. Generally, the code applies to all occupants regardless of their physical or mental abilities.

To provide for the safety of people and property involved in a building fire and to provide the facilities for firefighting and rescue operations, it is imperative that the design community place elements into the structure that will mitigate the growth potential of a fire. This section indicates the need for the building to have safeguards designed into it to achieve the objectives related to the protection of people and property. Specifically, the code states that a person not directly adjacent to or involved in the ignition of a fire must not suffer serious injury or death. It goes on to state that property loss has an upper limit based on the performance group assignment from Chapter 3. This section therefore establishes a single performance level with regard to occupant and public safety and an upper limit on the levels of damage with regard to property protection dependent on the performance group.

This particular functional statement is linked to Chapter 3 for guidance on the levels of performance. Because fire is an event whose magnitude is dependent on many factors related to the use, construction, configuration, size and contents of a building or facility, evaluating fire based on a simplistic relationship between magnitude of event and level of damage was not only inappropriate but extremely difficult; therefore, the requirements state "consistent with the performance group determined in Chapter 3." The general relationship between magnitude of event and level of damage presented in Chapter 3 is not completely invalid for fire, but the many interdependencies related to fire events must be specifically accounted for and understood. It should be noted that the more simplistic relationship between event magnitude and level of damage is more easily applied to seismic events or wind events because a building or facility will have no effect on the size (magnitude) of an earthquake or hurricane.

In addition to the concern with dependent factors, there was a concern that the levels of impact that were generically provided for all events, specifically life safety, were not necessarily appropriate for events such as fire. Society has a very low tolerance for death or serious injury caused by fires, especially of large numbers in a single incident. In addressing this low tolerance, this section in effect establishes a single performance level with regard to life safety in buildings, which basically equates to a mild impact for all fire events.

This single performance level recognizes that it is very difficult to protect someone who is intimately involved with the fire or in the immediate vicinity of the ignition point or source. Society does, however, have a higher tolerance for property loss, taking into account the importance of the building, which is why the upper limit of damage to be tolerated is based on the performance group classification of the building. Clearly the approach has been to separate property protection from life safety, which are two distinct objectives of building and fire codes. The performance level chosen is ultimately a public policy decision that reflects the expectations of society. Such expectations fluctuate in light of certain events, such as hurricanes, earthquakes and terrorist attacks such as those on the World Trade Center. In particular, the World Trade Center tower collapses resulting from the attack on 9/11 sparked considerable discussion and debate over the expectations of society for building construction.

The discussion of the events of 9/11 has generally never quantified to what level a building should be protected from such events, but the resulting recommendations from the National Institute of Standards and Technology WTC study have targeted specific issues within building and fire codes. It is not believed that buildings in general be protected from extreme events such as those of 9/11, but the study did reveal areas that can enhance the safety in buildings such as high-rise buildings, particularly very tall high-rise buildings that must self-sustain due to their height during an emergency. Also, very tall buildings are those more likely to be iconic in stature and warrant additional protection. There have been several changes specific to high-rise buildings in the IBC in recent years that relate to these enhancements. They primarily relate to structural integrity during fire, egress and firefighter safety. More specifically, in buildings over 420 feet (128 m) in height, fire-resistance reductions typically allowed for all high-rise buildings are no longer allowed. Additionally, reductions in column protection is not allowed in any high-rise building. Another structural fire protection enhancement relates to increased requirements for the bond strength of sprayed fire-resistance materials for high-rise buildings. In terms of egress, an additional stair has been added to buildings over 420 feet (128 m) in building height intended to address concerns of "counterflow." Counterflow is the issue of egress and fire department activities interfering with one another. Another issue related to egress is the use of elevators for egress, which in the IBC is explicitly allowed under certain conditions. In terms of firefighter safety, buildings with an occupied floor over 120 feet (36.6 m) from the lowest level of fire department vehicle access require two fire service access elevators (FSAE). This elevator is a part of a package of requirements including elements such as protected lobbies and stairs adjacent to the lobby (or a protected path to the stairway) that contain standpipes. All of these elements serve to increase the baseline of safety expected from prescriptive building and fire codes. Generally, a performance code already addresses these elements in the form of objectives, functional statements and performance requirements. The prescriptive code simply creates one of the solutions, which tends to provide a baseline for design.

Please note that a resource focused on fire protection for very tall buildings has been jointly developed by the Society of Fire Protection Engineers and the ICC. The document, now in its second edition, is titled *Engineering Guide—Fire Safety For Very Tall Buildings* (2021). The document provides a performance-based perspective of issues that are specific to very tall (high-rise) buildings.

As noted numerous times within this User's Guide, it is very difficult to understand the specific level of performance provided by the prescriptive code because the prescriptive code focuses on one or a limited number of solutions. The *ICC Performance Code® for Buildings and Facilities* has attempted to capture the intent of the prescriptive code.

1502.3 Performance requirements

The Fire Safety Concepts Tree, as published in NFPA 550, implies that fire safety requires either the prevention of a fire or the management of the impact of a fire that does occur. Managing the impact of a fire involves managing both the fire and the occupants of the facility that are exposed to that fire. In this document, managing the exposed occupants is addressed in Section 601, so those performance requirements are found there. The performance requirements in Section 1502.3 address the other strategies of controlling combustion or controlling fire spread. The general performance statement first makes a link back to the performance level stated in the functional statement by declaring, "Facilities or portions thereof shall be designed, constructed and operated to normally prevent any fire from growing to a stage that would cause life loss or serious injury." This section also states that should a facility sustain local fire damage it must remain intact as a whole, and not be damaged to an extent disproportionate to the local damage. For instance, a small wastebasket fire in a large open warehouse should not impair the structural integrity of the building or cause significant damage to the stored commodities. The remaining performance requirements refer to the more detailed elements of the building or facility, such as interior finishes, concealed spaces and vertical openings.

Fire management includes both active and passive fire protection strategies. Passive strategies include compartmentation (using fire-resistive materials) and structural fire resistance, while active strategies include fire detection systems, automatic fire suppression systems and manual firefighting. In this document, the provisions addressing manual firefighting by emergency response personnel are covered in Section 1502.3.3, as well as in Chapter 17, Emergency Access and Facilities, and Chapter 18, Emergency Responder Safety.

A fire can be managed by controlling the combustion process and the fire loading to restrict the size and rate of growth of the fire, or it can be controlled by limiting the spread of fire through the building and to adjacent buildings using active or passive means. Fire management can also be accomplished by a combination of the two approaches.

Controlling the combustion process and the fire environment involves limiting, where possible, the fuel loading in the facility to reduce the potential size and growth rate of a fire that does occur. Prescriptive fire codes have usually dealt with the concept of managing fire impact by controlling the fuel loading within buildings for special hazards. An example of such provisions is the maximum allowable quantities of flammable liquids allowed in nonhazardous occupancies (Chapters 3 and 4 of the IBC and Chapters 50 and 57 of the IFC.

Limiting the spread of fire and products of combustion through a building is also a function of managing the fire. This can be accomplished through passive methods such as providing fire-rated compartment barriers and opening protection, smoke barriers, shaft enclosures and fire-resistance-rated construction isolating higher hazard areas. Active systems, such as automatic sprinkler systems, other fire extinguishing systems and smoke management systems may also be relied on to prevent the spread of fire and products of combustion throughout the building.

Fire detection and alarm systems may also be considered as a means to initiate efforts to limit the spread of fire through the building, either by automatic activation, such as activation of smoke control systems or HVAC system shutdown, or by providing notification for initiating manual mitigation efforts by responsible employees or the public fire department. As indicated above, issues related to the notification, response and operation of emergency responder personnel are covered in other areas of this document.

Managing the fire impact also applies to reducing the impact of a fire on adjacent properties, processes and facilities. Features such as fire-resistance-rated exterior walls and exterior opening protectives, barriers to radiant heat exposure and water spray systems are intended to reduce the impact to these exposed facilities.

In addition to limiting fire spread to adjacent buildings, managing the fire impact also includes limiting the spread of fire to the building in question from an exposure fire. The use of fuel modification zones and special construction methods in wildland fire areas are good examples of this strategy. Providing adequate separation distances to aboveground fuel storage tanks is another tactic.

As stated above, a variety of strategies can be employed to manage the impact of a fire. Additionally, varying degrees of performance can be achieved. The necessary performance depends on many factors, such as building use and construction characteristics, occupant characteristics, location of the building, etc. As noted earlier, the functional statement establishes a single level of performance for fire life safety and provides upper limits for property loss, depending on the performance group designation of the building. To assist in the analysis, the magnitudes of possible fire events need to be established based on the fire load present and its impact evaluated against the specified level of performance.

Section 1502.3.15 and associated subsections specifically focus on the determination of the magnitude of events (i.e., fire scenarios) to be used in the analysis. This section requires the design fire events to realistically reflect the ignition, growth and spread potential of fires and fire effluents. Unlike for a natural hazard event, many elements of the building and its contents must be taken into account when developing fire scenarios to be used in the analysis. This section provides specific guidance on how the analysis is to be undertaken with regard to ignitability, heat release rate and overall fuel load. A key requirement is that the analysis must specifically look at a range of fire sizes to ensure that a small fire event does not create an unreasonable amount of damage and that a very large fire event satisfies the limits on the extent of damage based on the performance group. In no case should there be any anticipated loss of life or serious injury to persons not intimate with the initiation of the fire event, regardless of its magnitude.

The analysis with regard to fire will involve a single performance level for life safety and upper limits on property loss based on the performance group designation.

Fire-safety engineering and fire science currently involve a significant amount of uncertainty due to the lack of knowledge and accurate data in certain areas. Therefore, similar to the requirements of Chapter 5 for structural stability, one of the performance requirements specifies that design fires and fire scenarios be chosen that provide appropriate factors of safety or a given degree of redundancy between active and passive fire protection strategies. Addressing issues such as terrorism, in the wake of the 9/11 World Trade Center disaster, makes such uncertainty greater. The prescriptive codes have not typically addressed such events; therefore, a specific decision to address such scenarios would need to be made either by the community or by building owner. This code establishes a framework to factor in such issues.

Acceptable methods

Sources of ignition. As noted previously, the prescriptive code documents may include the IFC, IMC, the IFGC and the *National Electrical Code*. The listing requirements and manufacturer's instructions provide an additional set of prescriptive guidance documents to achieve the objectives.

Management of fire impact. As discussed earlier, the IBC deals with the management of fire impact in many areas including, but not limited to, Chapters 4, 5, 6, 7, 8, 9, 14 and 26. Other I-Codes have a role in managing fire impact, including the IFC, IEBC and *International Residential Code*® (IRC®).

CHAPTER 16 MANAGEMENT OF PEOPLE

This chapter addresses the role people might perform in a performance design. Many times, hazardous materials facilities depend on the building occupants and users to perform certain tasks to avoid and mitigate emergencies. These activities are considered integral to the success of such facilities. Other examples include restrictions on the types of appliances allowed in the lunchroom or the number of staff available for certain types of events such as sporting events. This section outlines not only the more traditional prevention and protection skills that the existing prescriptive codes require but also requires that where the actions or practices of people become a component of a design, they must be maintained.

Functional statements in Sections 1601.2.1 and 1601.2.2 indicate to the code user that this chapter should be used in conjunction with Chapter 3 to establish the required design performance level. Performance requirements in Sections 1601.3.1 through 1601.3.10 list specific criteria that must be addressed after determining the magnitude of design events and while completing the hazard analysis to ensure that the appropriate level of performance and protection is achieved.

The overall objective as well as each of the specific performance requirements refers to "people," "occupants" or "staff." The purpose is to clearly separate the performance of people from the performance of building features or systems. The concept that this is more than an education and training issue but is critical to the system reliability was felt to be important. As an example, the actions of people as opposed to the redundancy of systems may be used as a method to provide reliability. Or, the availability of a highly trained on-site emergency response unit (plant fire or hazmat brigade) rather than built-in systems or other process features may be used as a design or solution alternative. In either case, the actions of people become a component of the design or solution and are as important to the overall outcome as the more traditional built-in features of the design, process or solution. This particular document does not intend to promote designs utilizing people as a design solution in the place of systems such as sprinklers, but it does emphasize that the role people play in preventing and managing emergencies should not be ignored.

Performance designs, in some cases, may depend more heavily on the actions of occupants than a traditional prescriptive code design. The prescriptive code has traditionally required that some occupants and staff receive training and/or drills (students in schools or staff in nursing homes and hospitals). Very rarely has the fire code required that occupants or staff be relied on for critical functions (with some exceptions for hazardous materials). Because the actions of people are an important safety function, specific requirements for education both to promote safe practices and actions and to develop specific procedures and training are found within the performance requirements.

Because occupants, staff, equipment, materials and processes can change, it was felt that requiring the establishment of some administrative controls in order to maintain consistency in the level of knowledge was important. The loss of certain staff can change the overall competency of the staff to react to and prepare for emergencies, especially if the staff lost served in a leadership role in such activities. A change in the products that are stored or the introduction of a new technology into an existing process can also impact the original design or solution. The administrative controls are meant to require ongoing evaluation and validation so that the original assumptions remain valid. They are also meant to require appropriate education and training whenever new people or other original assumptions change. Such ongoing evaluations, education and training are also required in Chapter 4 of the *International Fire Code*® (IFC®).

The action or the inaction of people can be as key to the design or performance solution as any built-in system or feature.

The extent of the actions of building or facility occupants or users will depend on the design chosen. More emphasis may be placed on an automatic system, for example, with lesser dependence on the actions of building or facility occupants.

Acceptable solution

Examples of code provisions focused upon people are found throughout the IFC. This is especially true of Chapter 4 of the IFC which addresses emergency planning and preparedness. An example focused upon hazardous activities within the IFC could include aspects such as the requirements or attendants at motor-fuel dispensing stations, a fire watch for welding operations or the housekeeping requirements for high-piled storage facilities.

CHAPTER

17 EMERGENCY ACCESS AND FACILITIES

Inevitably, various emergencies, including medical emergencies, occur at buildings and facilities. This chapter deals with providing suitable access and facilities for emergency operations and responders. The intent of this chapter is to address the need for some manual action to preserve the safety of people and limit damage to a building or structure, and its contents.

Some of the conditions that might warrant emergency response and action include fires, storms (tornadoes, hurricanes, severe thunderstorms), bomb threats, hazardous materials releases, hazardous conditions from utilities, medical emergencies or any other condition that can lead to injury or property damage. To achieve these objectives, two components were felt to be important: access and facilities.

Chapter 18 specifically addresses emergency responder safety. This chapter focuses on the tools needed to appropriately undertake emergency response.

Exterior access and facilities

The emergency response capability of communities varies greatly from jurisdiction to jurisdiction. The complement of apparatus and staffing that might respond to an emergency will vary by jurisdiction and by the reported type of emergency incident.

In such instances, it is essential that emergency response units and crews be able to rapidly access the property or building. Additionally, units and crews must both be able to reach and utilize facilities provided about or within the property to specifically aid emergency responders and facilitate management of an emergency at or within a property.

Accordingly, designers of buildings and facilities must communicate with the appropriate public safety officials to ascertain what will happen when the inevitable emergencies occur at or within the property. The needs and expectations of those officials and the capabilities of the emergency response forces that will be called upon to manage or assist in the management of the incident must be incorporated into the design and standard operating procedures for the property. Agreed-upon methodologies to accomplish fire safety objectives will be incorporated into the approved design, and the mutually established procedures must be incorporated into both the property's "owner's manual" and the public safety agencies' operating procedures. A design that depends heavily upon emergency response must provide for changing capabilities of the emergency responders over time, which may be very difficult. This chapter establishes performance criteria for (1) access to properties by the units/apparatus that transport the emergency response staff; (2) the deployment of units/apparatus and equipment utilized by the emergency response forces in responding to and managing the emergency; and (3) the staging of apparatus, equipment, staff and facilities that might ultimately need to be deployed or utilized in management of an emergency at a property or building.

Emergency responders usually respond to emergencies on specialized apparatus/units designed and equipped to support personnel in management of an emergency. As an example, urban communities will usually provide aerial ladder (or tower ladder) apparatus equipped with specially designed, hydraulically operated, multisectioned ladders with a vertical reach of 100 feet (30 480 mm) or more to enable firefighters to effect rescues from several stories above grade level or operate water streams into upper portions of low-rise buildings. Rural communities may provide water tenders or tankers with 2,000 gallons (7571 L) or more of water designed to quickly dump the water into fold-up tanks so that, oftentimes, in conjunction with multiple tenders, water can be shuttled from a source (lake or pond) to the fire scene. Typically, all fire apparatus carry large complements of tools and equipment that are removed and carried by emergency responders to various locations around or within the facility to aid in the management of the emergency.

Most emergency situations will involve the response of several emergency vehicles or apparatus. It may be necessary for some of those vehicles to leave and return to the incident scene one or more times in the course of the management of the emergency. For example, a water tender may make many trips back and forth from the water supply source to the scene of a fire to provide the water necessary to meet the fire-flow demand. In site planning it is important for designers to carefully research and consider the potential emergency services response complements and the type(s) of apparatus that will respond to the facility. Designers must ensure that those units will be able to adequately access the site or facility and take the tactical positions that provide optimum advantage of the capabilities of the emergency response complement. To accomplish this, designers must consider such things as the number, type, size (length, width and height), weight [gross vehicle weight (GVW)], turning radius, operational features (width of an aerial apparatus with stabilizing jacks deployed), and tool and equipment inventory of the apparatus complement that reasonably can be expected to respond and operate at a major emergency at the site or facility.

Accordingly, clearances must be provided that allow the ingress and egress of emergency vehicles. There must be sufficient access and space for those vehicles to stage, park, pass and deploy in tactically advantageous locations and positions. Apparatus carrying special equipment and tools that will be carried into or deployed around the structure must be able to park near entrances or deployment points. The roadways, lanes, staging areas and parking surfaces must be able to carry the load imposed by the specialized apparatus used by the emergency response forces and be traversable in all seasons or reasonably expected weather conditions. Care must be taken to ensure that appurtenances and protrusions from buildings and structures will not impede access and that, as necessary for life-safety, accommodation is made to ensure that aerial apparatus can deploy to reach the maximum height and sweep of a building side or face.

Grades and slopes must accommodate the apparatus and facilitate access by emergency response forces and deployment of their equipment; i.e., the placement of ground ladders. Once parked at the scene of an emergency, emergency response forces will deploy and utilize equipment brought to the scene on large apparatus, and equipment and systems installed or provided at the facility, in the management of the emergency. In the early moments after arrival, emergency responders will be concerned with establishing a supply of water to sustain an attack on the fire (usually by connecting hose lines from a fire hydrant to a fire pumper) and stretching or extending hoselines from pumpers or standpipe systems to the area of the fire. Such operations require the commitment of a substantial number of emergency responders and take valuable time in the early moments after arrival when many priority tasks must be completed in as short a time as possible.

Interior access and facilities

Once the exterior staging is underway, interior operations become important for emergency responders. There might be many instances where an exterior attack is all that is required or possible, based on the hazard present. The requirements relating to interior access and facilities address such buildings and facilities where it is reasonable to expect that the emergency responders will need to mitigate a hazard from the interior. Some of the main issues addressed in this chapter in this regard include the interaction of the occupants' egress and emergency responders' access; necessary equipment and building or facility layout; and for larger, taller buildings, features such as elevators.

A performance requirement that specifically deals with the interaction of the means of egress and the emergency responder access is included to ensure that the two functions do not conflict with one another. This conflict is often termed "counterflow" and was an area of interest during the World Trade Center study. In many cases the occupants will likely egress before the fire department arrives, but in cases such as high-rise buildings, the egress process can take much longer, depending upon the egress strategy. Also, if there are evacuation strategies for buildings such as high rises that would call for full building evacuation, this problem may be more prevalent depending upon where the emergency responders are located within the building. The SFPE/ICC *Engineering Guide for Very Tall Buildings* discusses various egress strategies that might be applied. It would be unrealistic for certain building features such as a stairway to be available for both activities at the same time. As a result of this concern during the WTC study, the IBC requires an additional stair within high-rise buildings over 420 feet (128 m) in building height. This is a prescriptive solution to counterflow. There is an exception that obviates the need for this additional stair (Occupancy Evacuation Elevators provided for occupant egress that are protected by special protection including fire-resistance-rating lobbies and standby power).

The second aspect addressed with regard to interior access and facilities is related to providing appropriate equipment for emergency response such as standpipes, Self-contained breathing apparatus (SCBA) and staging areas. The level of need in this area is heavily dependent on the building itself and the needs of the emergency responders, related to their abilities. This kind of information is obtained through interaction with the emergency responders during the design of a building or facility.

Finally, in large buildings and very tall buildings, a means might be necessary to move emergency responders and their equipment vertically and, in some cases, horizontally within buildings. This usually involves the use of elevators for vertical movement. Self-propelled vehicles such as golf carts might be provided for facilitating horizontal movement. The prescriptive codes have requirements for elevator recall (Phase I), emergency operation (Phase II) and standby power. There is also an additional package of requirements related to the fire department's access and facilities needs for tall buildings in the *International Building Code*® (IBC®). More specifically, fire service access elevators (FSAEs) are required in buildings containing occupied floors more than 120 feet (36.6 m) above the lowest level of fire department vehicle access. Basically, the package of requirements builds on the current requirements for recall and emergency operation of the elevator and adds requirements for elevator lobbies, location and protection of stairs containing a standpipe connection adjacent to such lobbies and several other enhanced features.

CHAPTER 18 EMERGENCY RESPONDER SAFETY

This chapter is unique because it not only attempts to bring together the major issues impacting emergency responder safety found throughout prescriptive codes but also provides added importance to this issue by making it a separate section. The prescriptive codes have always made provisions to lessen the dangers to emergency responders, firefighters in particular; however, this code takes the next step and places these concerns into a separate chapter. This does not mean that many of the provisions found elsewhere in this document do not also pertain to firefighter safety but only that specific issues related thereto are addressed in this chapter. Examples of such other provisions can be found in the chapters on egress, access and hazardous materials.

This chapter was developed with the understanding that emergency response by its nature is inherently dangerous, and it is beyond the realm of possibility to remove all hazards faced by responders. However, much can be done to provide reasonable levels of safety. The scope has been limited to alleviating those hazards that are beyond what would normally be expected during an emergency. Additionally, it is vital that each jurisdiction give careful consideration to what is an acceptable risk for emergency responders and provides appropriate input during the design of the building, facilities and premises to establish the appropriate level of risk.

Section 1502.2.6, requires that buildings be designed to allow firefighters to perform the necessary tasks in the event of a fire.

The functional statements outlined in Items 1 through 3 of Section 1801.2 provide the basic guidance for the user as to what general areas need to be considered when evaluating provisions for emergency responder safety. The user would first establish the required performance level in conjunction with Chapter 3 after a thorough evaluation of the potential hazards to responders. Performance requirements detailed in Sections 1801.3.1 through 1801.3.4 list specific criteria that must be addressed after determining the magnitude of design events and while completing the hazard analysis to assure the user has achieved the appropriate level of performance and provided adequate protection for the responders depending on the hazards identified.

Four distinct performance requirements are included in this chapter that specifically address responder safety and functionality. The first two specify that hazards be clearly identified to the responders. This provision can be achieved by signage, barriers or other common means or might require an extensive database system available to responders. Such information might come from the designer in the form of facility layout and description of contents. This information would, of course, be based on the severity of the hazard and the specific needs of the responding agency.

The third functional statement addresses similar issues found in Section 1502 of this code but specifically intends to prevent unexpected structural failure. It was understood when these provisions were written that it is neither economically feasible nor realistic to design and build structures and facilities in such a manner that collapse or failure could be eliminated in every instance. Considering this, language is included to indicate that failures should be as predictable as possible considering the incident severity, construction materials and methods, and incident duration. This is more clearly spelled out in Chapter 15.

The final functional statement ensures that emergency responder communication needs are considered. Again, this becomes very specific to the needs of the local jurisdictions as well as the unique needs dictated by the building. This communication could be as simple as standard firefighter radios, phones and public address systems or as extensive as an in-building emergency responder communications enhancement system.

Acceptable methods

Construction in accordance with the *International Building Code*® (IBC®) and *International Fire Code*® (IFC®) would be deemed to comply with the requirements of Chapter 18. The original provisions of this chapter were written to reflect, as close as possible, what the prescriptive codes intended in terms of the protection of emergency responders. This includes minimum levels of construction, proper identification and signage of hazards and providing the ability to communicate during an emergency within the building.

In terms of communication, Section 510 of the IFC requires in-building emergency responder communication enhancements systems (ERCES) as applicable to the building and coverage provided. The IFC also triggers for existing buildings under certain circumstances. The IBC provides a trigger to evaluate the need for such systems for new construction within Chapter 9.

CHAPTER 19 HAZARDOUS MATERIALS

Although performance-based design approaches are relatively new to the building construction and fire safety arenas, such approaches have been in widespread use in the hazardous materials arena for quite some time. Numerous regulatory programs enacted by the federal government in the 1980s and 1990s encouraged or required the use of performance-based risk management techniques for many facilities and processes involving hazardous materials, and the documentation associated with these programs served as an excellent resource for use in the development of Chapter 19.

Users of Chapter 19 will note that the technical provisions from two such federal programs, Process Safety Management (PSM) and Risk Management Planning (RMP), served as the specific source for most of the performance requirements now found in Chapter 19. Those unfamiliar with these programs will note that PSM falls under the auspices of the U.S. Occupational Safety and Health Administration (OSHA) and RMP falls under the auspices of the U.S. Environmental Protection Agency (EPA). The source regulations can be found in Titles 29 and 40 of the Code Federal Regulations. The remaining performance requirements in Chapter 19 were developed to include topic areas currently covered by the *International Fire Code*® (IFC®) that were not reflected in the PSM or RMP rules and to include the reporting requirements set forth in SARA Title III. Overall, the performance requirements in Chapter 19 fully cover the regulatory topic areas for hazardous materials that are encompassed in the IFC.

When this chapter was originally written, the goal was to provide consistency with the existing PSM and RMP programs.

While closely tied in concept to the IFC and federal PSM and RMP rules, there is at least one major difference between Chapter 19 and these other documents. Chapter 19 does not make use of prescriptive, material- and quantity-based thresholds as a baseline for determining when the chapter applies. However, this should not be taken as an indication that the PSM- and RMP-type rules in Chapter 19 are mandatory in all cases.

Chapter 19 and the PSM- and RMP-type rules therein present an optional method of design. Where it is advantageous, the owner may prefer this method. Presumably, this would involve a facility that is already required to follow Federal PSM and/or RMP rules. Where it is not advantageous to an owner to use performance-based design methods, prescriptive codes are always available. Operators of small facilities that are exempt from federal hazardous materials regulations will likely opt to use the requirements in the prescriptive IFC and *International Building Code*® (IBC®) as they will tend to be less onerous for smaller facilities. Note that the *International Fire Code*® (IFC®) has specifically incorporated the language from this chapter into Chapter 50 of that code to provide this option.

SECTION 1901—HAZARDOUS MATERIALS

1901.1 Objective

The intent and scope of Chapter 19 is similar to the intent and scope of Chapter 50 of the IFC prescriptive code: to protect occupants of the building, people in the surrounding area, emergency response personnel and property from acute consequences associated with unintended or unauthorized releases of hazardous materials. Like the prescriptive IFC and IBC, the performance code encourages the use of both accident prevention and control measures to reduce risk.

It is not the intent of this code or the prescriptive codes to regulate all hazardous materials. Within the scopes of building and fire codes, hazardous materials are generally defined as those materials that are acutely dangerous to people or property. Building and fire codes usually defer regulation of materials that present only a risk of chronic or environmental effects to other regulatory agencies, such as OSHA or the EPA in the United States. Exposure of workers to hazardous materials in the normal course of their jobs is also beyond the scope of building and fire codes. Such workplace safety issues are instead regulated by occupational safety and health codes, which in the United States fall under the jurisdiction of OSHA.

Hazardous materials regulated by building and fire codes are typically classified into two major categories: physical hazards and health hazards. Physical hazard materials are those materials that present a risk of fire, explosion or accelerated burning, and include the following:

- Explosives and blasting agents.
- Flammable and combustible liquids.
- Flammable solids and gases.
- Organic peroxide materials.
- Oxidizing materials.
- Pyrophoric materials.
- Unstable (reactive) materials.
- Water reactive solids and liquids.

Health hazards materials are those materials that present a risk of acute health consequences from a short-term exposure, and include the following:

- Toxic and highly toxic materials.
- Corrosive materials.

When developing a performance-based design involving hazardous materials concerns, consideration should be given not only to the foregoing classifications but also to the quantity, state, situation (storage/use), arrangement, and location of materials and processes.

1901.2 Functional statements

Chapter 19 includes two functional statements that serve the overall objective of the chapter. These two statements focus on reducing the probability of unsafe conditions involving hazardous materials and minimizing the consequences of an unsafe condition, if one occurs. The concepts can be summarized as prevention and control. Specific means by which these functional statements can be accommodated are listed in the performance requirement section.

The code includes some additional functional statements that are relevant to hazardous materials but are not included in the text of Chapter 19. These include some of the functional statements found in Chapter 4, Reliability, and Chapter 16, Management of People.

1901.2.1 Prevention

This section invokes the need for consideration of accident-prevention techniques. Such techniques might include administrative policies and procedures specifying safe practices. Clearly, it is preferable from a safety standpoint to prevent accidents as opposed to dealing with consequences after a release or failure has occurred. Nevertheless, it is important to provide for an appropriate degree of consequence management as well because it is not reasonably possible to prevent all accidents. Hence, the need for the functional statement in Section 1901.2.2.

1901.2.2 Mitigation

Management of consequences can be accomplished in a variety of ways. The most effective methods are those that quickly detect and respond to abnormal conditions before severe consequences occur. Limit switches with integral process-shutdown capability are an example of such an approach. Other approaches to consequence management involve the use of protection methods to limit consequences once a spill or release has occurred. Spill containment systems, ventilation systems, fire sprinkler systems and fire-resistive construction methods are all examples of protection methods that limit consequences.

1901.3 Performance requirements

Recognizing that functional statements are deliberately broad in their effort to establish general direction, the code provides performance requirements as a means by which the principles embraced in functional statements can be accomplished. The principles embodied in the performance requirements set forth in the performance code are generally consistent with those embodied in prescriptive codes. However, performance-based design methods allow a more systems-oriented/risk-based approach because prescriptive codes do not generally recognize the beneficial interaction of various protection methods. Therefore, prescriptive design methods can result in unnecessary and inefficient redundancies in design, or in some cases hazards may not be sufficiently addressed. Performance-based design methods allow a more comprehensive approach that more clearly demonstrates the intended outcomes.

1901.3.1 Properties of hazardous materials

Section 1901.3.1 correlates with the reporting requirements set forth in SARA Title III and to some degree with the prescriptive reporting requirements set forth in some model fire codes. Compliance with these reporting requirements can be accomplished through the use of safety data sheets (SDSs), inventory reports, SARA Title III reporting documents (which are typically mandatory under federal law), etc.

This section ensures that interested parties will have access to information about the characteristics of hazardous materials that are located on site.

1901.3.2 Reliability of equipment and operations

Equipment and operations at facilities regulated by federal PSM rules should have little trouble demonstrating compliance with the requirements of Section 1901.3.2. The PSM rules generally address this topic area.

At facilities that are not required to comply with PSM rules, the selection of equipment and design of operations would have to go through a great deal of scrutiny by qualified individuals. In addition, equipment manuals and operational protocols would need to be developed and followed, as applicable.

1901.3.3 Prevention of unintentional reaction or release

Facilities regulated by RMP rules are required to evaluate the potential consequences of various release scenarios on the surrounding area, and therefore many such facilities provide safety systems to reduce these potential consequences, recognizing that the consequence analysis information must be made available to the public.

Depending on the classification and state (solid, liquid or gas) of hazardous materials stored or used at a given site, a variety of mitigation measures may be provided to comply with this provision. Such measures might include process controls, spill control and containment systems, and ventilation controls.

1901.3.4 Spill mitigation

This requirement is primarily derived from provisions in the prescriptive IFC. As a general rule, storage facilities are regarded as less likely candidates for dangerous spills than facilities that involve dispensing or processing operations. In addition, dangerous spill conditions are probably more likely to occur in facilities with large quantity vessels or systems than those with only small containers. Information that may be useful in determining whether a spill is plausible and whether dangerous conditions would result include the following:

- Specific material and process hazards involved.
- A block flow diagram for the facility.
- Piping and instrument drawings.
- A list of all safety devices, showing their location, design basis and capacity, date of installation, etc.
- Equipment manufacturers' operational instructions, including safe operating limits for the equipment.
- Equipment drawings and specifications that reflect built and installed equipment.

1901.3.5 Ignition hazards

The primary design and operating intent is to ensure that flammable and combustible materials are always completely controlled in accordance with process design parameters. However, where flammable and combustible hazardous materials are present, a degree of redundancy is sometimes necessary to provide an additional level of safety. Where there is a plausible risk of spills or leaks, such as in loading and unloading or packaging operations, additional measures such as ignition source controls are prudent. To that end, process design and operation should ensure to the greatest degree possible that ignition sources are kept away from areas where flammable or combustible hazardous materials are present. Where separation is not feasible, ignition source controls may be warranted. Such controls may involve the following:

- Electrical classification of areas where flammable hazardous materials might be present.
- Classification of mobile equipment that might operate in areas where flammable hazardous materials might be present.
- The use of grounding systems and equipment to minimize the potential for sparking in areas where flammable hazardous materials might be present.

1901.3.6 Protection of hazardous materials

This section directs the designer and the operator to review and ensure that vessels or systems containing hazardous materials are not exposed to or are protected from damage by external fire. The design should focus first on reducing the possibility for fire or other hazards, such as vehicular impact, and second on isolating hazardous materials from exposure to unsafe conditions, such as a fire.

All storage areas and systems should be formally reviewed to find and correct any sources of exposure to fire, including the following:

- Nearby storage of combustibles.
- Nearby hot-work operation.
- Nearby vehicular operation.

All systems subject to fire exposure should be formally reviewed to ensure adequate protection, including the following:

- Sprinkler installation.
- Insulation of equipment.
- Fire-resistive barriers.

1901.3.7 Exposure hazards

This section directs the designer and the operator to review and ensure that vessels or systems containing hazardous materials are not subject to damage from internal fire, chemical reaction or explosion. The design criteria should be first to reduce the risk of an internal fire or explosion, and second, where the first is not feasible, to design vessels and systems in such a manner that loss of integrity will not occur in an overpressure situation.

All systems should be formally reviewed to identify and correct any sources of internal fire, explosion or overpressure. The review should include the following:

- The potential for inadvertent or improper mixing of reactive components.
- The potential for overheating of unstable materials.

- The potential for inadequate venting of unstable reaction byproducts.
- The potential for inadequate dilutent material supply.

Where overpressure or explosion conditions cannot be reasonably ruled out, the design should consider overpressure protection, containment and explosion control systems.

1901.3.8 Detection of gas or vapor release

This section is derived from the IFC. The section ensures that hazardous vapor releases are detected and mitigated before they can harm individuals or property. In occupied areas, detection of a vapor release may be by sight, smell or an automatic detection system. For many hazardous materials such as chlorine or ammonia, vapor releases are readily evident before concentrations are truly hazardous based on the presence of vapor fog or a noxious odor. Where this is not the case, automatic detection systems and alarms may be warranted. Sensors can take the form of ambient sampling devices at strategic area locations, sampling devices in key vent streams or specially designed leak-detection systems such as acoustic emission systems. The performance measurement is the ability of the sensing equipment or operators to provide adequate warning so that safety precautions can be taken before unsafe conditions are present.

Mitigation-based solutions can range from special process equipment designs to elaborate ventilation and air scrubbing systems. Where practical, the simplest mitigation consists of over-design of the process system so that the likelihood of release is extremely low. The performance measurement of a ventilation or treatment system is the reduction of the concentration of the hazardous material in the workplace and nearby environment to levels that are not acutely hazardous.

1901.3.9 Reliable power source

This section is derived from the IFC. It is essential to safety to ensure that a reliable power supply is provided for systems that are critical to safety. Some examples of systems that may require a reliable power supply include mechanical ventilation systems, treatment systems, gas detection and alarm systems, and emergency shutdown systems. The reliability needs of the system are related to the potential risks associated with system failure.

A reliable power source does not necessarily equate to a generator or a battery system. The type of system to be used depends on the relative level of hazard that might result in the event of a power failure, and in some cases, such as those where hazardous processes shut down upon loss of power, a connection ahead of the building main disconnect switch may be adequate to qualify as a reliable source. Guidance on the selection and performance requirements for power supply systems providing an alternate source of electrical power can be found in the *National Electrical Code*; NFPA Standard 110, *Standard for Emergency and Standby Power Systems*; and NFPA 111, *Standard on Stored Electrical Energy Emergency and Standby Power Systems*.

1901.3.10 Ventilation

In many cases involving hazardous materials, ventilation must be provided to limit the risk of creating an emergency condition. Ventilation might be necessary during both normal and abnormal operating conditions. Some examples of operations that may require ventilation are storage or processing of flammable and combustible liquids or gases inside of buildings, drum filling operations inside of buildings, laboratory use of chemicals and dust handling systems. Ventilation may also be used as a means for reducing vapor concentrations below lower flammable limits in areas where ignition sources are present or for pressurization of areas to isolate hazardous vapors.

Guidance on the performance requirements for ventilation systems can be found in a number of sources, including OSHA 29 CFR 1910.106, Flammable and Combustible Liquids; NFPA 30, Flammable and Combustible Liquids Code; NFPA 69, Explosion Prevention Systems; NFPA 45, Laboratories Using Chemicals; NFPA 70, *National Electrical Code*; and NFPA 497, Recommended Practice for Classification of Flammable Liquids, Gases or Vapors and of Hazardous (Classified) Locations for Electrical Installations in Chemical Process Areas.

1901.3.11 Process hazard analyses

This section establishes an administrative safety control addressing process hazard analysis. Guidance on process hazard analysis techniques can be found in the OSHA PSM regulations, 29 CFR Part 1910.119. The process hazard analysis must be appropriate to the complexity of the process and must identify, evaluate and control the hazards involved in the process. The analysis can be accomplished through various methods. Some of these are "What if," Process Hazard Analysis, HAZOP, fault tree, etc. A person trained in these and other hazard evaluation techniques should be employed to complete this analysis.

1901.3.12 Written procedures and enforcement for prestartup safety review

This section establishes an administrative safety control addressing prestartup safety review procedures. Guidance on techniques for written documentation of prestartup safety review procedures can be found in the OSHA PSM regulations, 29 CFR Part 1910.119. Prestartup safety reviews are typically necessary when new facilities are prepared for operation and where existing facilities are modified to a degree that is significant enough to require a change in the process safety information.

A prestartup safety review should confirm that prior to the introduction of highly hazardous chemicals to a process, the following verifications have been accomplished at a minimum:

- Construction and equipment is in accordance with design specifications.
- Safety, operating, maintenance and emergency procedures are in place and are adequate.
- For new facilities, a process hazard analysis has been performed and recommendations have been resolved or implemented before startup; for modified facilities, requirements contained in management of change documents have been met.
- Training of each employee involved in operating a process has been completed.

1901.3.13 Written procedures and enforcement for operation and emergency shutdown

This section establishes an administrative safety control addressing written documentation of operating procedures and emergency shutdown procedures. Guidance on developing written documentation for operating procedures and emergency shutdown techniques can be found in the OSHA PSM regulations, 29 CFR Part 1910.119. Overall, there are 14 elements that employers covered by PSM are required to complete to meet the Federal PSM regulations. Two elements that relate to this section are as follows:

- 29 CFR 1910.119 (c): This element requires that employees and their representatives be consulted on the development and conduct of hazard assessments and the development of chemical accident prevention plans and provide access to these and other records required under the federal law.
- 29 CFR 1910.119 (f): This element requires that written operating procedures for the chemical process including procedures for each operating phase, operating limitations, and safety and health considerations must be developed and implemented.

1901.3.14 Written procedures and enforcement for management of change

This section establishes an administrative safety control addressing management of change. Guidance on developing written documentation for management of change can be found in the OSHA PSM regulations, 29 CFR Part 1910.119. The PSM element that relates to this section is as follows:

29 CFR 1910.119 (l): This element requires a review of the technical basis for the proposed change; the impact of change on safety and health; possible modifications to operating procedures and process safety information; the necessary time period for the change; and authorization requirements for the proposed change.

Employees involved in operating a process and maintenance and contract employees whose job tasks will be affected by a change in the process should be informed of and trained in the change prior to startup of the process or affected part of the process.

1901.3.15 Written procedures for action in the event of emergency

This section establishes an administrative safety control addressing emergency response planning. Guidance on developing written documentation for an emergency response plan can be found in the OSHA PSM regulations, 29 CFR Part 1910.119. The PSM element that relates to this section is 29 CFR 1910.119 (n), which references other portions of the federal regulations. Such plans may include identification of actions to be taken by employees in the event of an emergency and the assignment of a staff liaison who can assist emergency response personnel.

1901.3.16 Written procedures for investigation and documentation of accidents

This section establishes an administrative safety control addressing accident investigation and reporting. Guidance on accident investigation and reporting can be found in the OSHA PSM regulations, 29 CFR Part 1910.119. The PSM element that relates to this section is 29 CFR 1910.119 (m).

Some of the guidelines specified in the federal regulations include the following:

- The need for an incident investigation team to be established, consisting of at least one person knowledgeable in the process involved, a contract employee if the incident involved work of the contractor, and other persons with appropriate knowledge and experience to thoroughly investigate and analyze the incident.
- The need for a report to be prepared at the conclusion of each investigation, including at a minimum: date of incident; date investigation began; description of the incident; factors that contributed to the incident; and recommendations resulting from the investigation.
- The need for establishment of a system to promptly address and resolve the incident report findings and recommendations, and to document resolutions and corrective actions.
- The need for accident investigation reports to be reviewed by all affected persons whose job tasks are relevant to the incident findings, including contract employees where applicable.

1901.3.17 Consequence analysis

This section establishes an administrative safety control addressing an analysis of off-site consequences. Guidance on accident investigation and reporting can be found in the EPA RMP regulations, 40 CFR Part 68. These regulations amend the accident release prevention requirements under Section 112 (r) of the Clean Air Act.

EPA's RMP rules are a good source of examples for alternate release scenarios for a particular site, and, through the identification and analysis of plausible release scenarios, changes can be implemented to minimize the probability and consequences of a release. A plausible release is one that has occurred in the past or could occur under reasonable single-system failures.

Devices that normally use some kind of motion or energy to prevent or minimize the release represent active mitigation controls. Active mitigation controls might include valves, switches, pumps and blowers. Passive mitigation controls include devices that are permanently in place and have an inherently safe design that allows them to be in effect at all times. Passive mitigation controls might include dikes, walls, ponds and sumps.

The off-site consequence analysis can be accomplished through various methods. Some of these are "What If," Process Hazard Analysis, HAZOP and fault tree. A person trained in these and other hazard evaluation techniques should be used to complete this analysis.

1901.3.18 Safety audits

This section establishes an administrative safety control addressing safety compliance audits.

Guidance on safety audits can be found in the OSHA PSM regulations, 29 CFR Part 1910.119.

The PSM element that relates to this section is 29 CFR 1910.119 (o).

On a routine basis, each facility must review its continuing compliance with each of the sections in Chapter 19 and other related chapters. The word "periodic" reflects a need for adequate frequency to have reasonable confidence that safety programs, features and systems will perform as intended. Recognizing that many code sections contain issues that change very little over time, compliance audit frequencies will not be the same for all programs features and systems, and depending on the particular safety element, the audit frequency may range to as much as three-year intervals under the PSM regulations.

1901.3.19 Levels of impact or damage

Each facility must satisfy the required design performance levels that are applicable to hazardous materials. These performance levels are defined in Section 304 as follows: mild, moderate, high and severe. These definitions help in standardizing performance-based design approaches. The intent of assigning design performance levels is to define the people and area impacted by an unwanted release of hazardous materials and the degree to which people and property are impacted.

With respect to hazardous materials, a mild impact event might equate to an incident that has a minor effect or damage on the people and property immediately involved in the work activity.

A moderate impact event might result from an incident that involves a moderate spill and/or fire that has significant local consequences to the people and property in the immediate area of the work activity and within logical containment boundaries around the area. This area might include a room or small warehouse.

A high impact event might result from a large spill of noncombustible materials, a room and contents fire, or a small deflagration. Such an event would be contained within the boundaries of the facility and would affect only the people and property within that boundary.

A severe impact event might result from a very large spill of noncombustible materials, a fire that extends beyond primary containment boundaries, or a large deflagration or detonation. Such an event would not be contained within the boundaries of the facility and would affect people and property outside that boundary, perhaps including the general public.

Life-safety and property protection can be looked at somewhat independently. Society generally has a lower tolerance for risks related to life-safety than it does for property loss. This should be kept in mind when addressing the level of impact or damage that can be tolerated.

Magnitudes of design events are specified as a means of indicating size of an incident that must be handled by a building or facility. Some examples of various sizes of events include a small spill, a localized fire or a large explosion or hazardous materials release. Magnitudes of design events relate to performance groups in that magnitudes specify the "threat" or "size" of an initiating incident, and performance groups limit the tolerable level of damage that might occur as a result of such incidents.

Sections 1901.3.19.1 through 1901.3.19.7 provide guidance on choosing the events to analyze. This guidance includes a review of factors to be addressed such as the use of the room or area, occupant risks and importance to the community. Four scenario sizes must be defined: small, medium, large and very large. As with fire, hazardous materials events are not events independent of the building's or facility's contents, layout, construction or its occupants. Therefore, these factors need to be addressed when determining the scenarios. This process must include a thorough engineering analysis with proper justification.

Acceptable methods

The acceptable methods that are essentially deemed to comply for these provisions can be found in Chapters 3 and 4 of the IBC and Chapters 50-67 of the IFC. Note that Chapters 20-41 of the IFC also address many special uses that involve the use of hazardous materials such as flammable finishing, motor vehicle fueling and repair, and combustible dust production.

APPENDIX

A RISK FACTORS OF USE AND OCCUPANCY CLASSIFICATIONS

A101 Objective

This section classifies buildings, structures and portions of buildings and structures by their primary use in order to facilitate design and construction in accordance with other provisions of this code. When determining the design performance level, the building or structure needs to be assigned to a performance group. This appendix provides guidance as to the use group or occupancy classification of the building to determine the performance group when applying Table 303.1. The performance group classifications are needed to address the language in the various provisions that requires design features to be appropriate to the use for performance-based acceptable solutions and to provide the basis for the use of applicable provisions of the *International Building Code*® (IBC®) as "deemed to comply" acceptable methods. If this appendix is used, any assumptions should be documented and verified. The descriptions found in this appendix need to be taken as a starting point, as many buildings and facilities have unique characteristics that would make a performance-based analysis on a general use and occupancy classification inappropriate.

A102 Functional statements

This section states which factors should be considered when determining the primary use of a building. These factors include the use of the occupancy classifications found in the IBC and an analysis of the hazards and risks to the occupants. It is again emphasized that the occupancy classifications alone may not be sufficient, because of the unique hazards and risks associated with a particular building or facility.

A103 Use and occupancy classifications

To provide consistency between the provisions of this code and the provisions of the IBC, the fundamental definitions for use and occupancy classifications are the same.

Should the IBC be used as the basis for an acceptable method for any of the use and occupancy classifications listed in this section, all of the applicable provisions of the IBC shall be applied.

A103.1 General

To provide a better understanding of the reasons and assumptions behind various use and occupancy classifications, and to provide a codified foundation for use of performance-based analysis and design methods as the basis for developing acceptable solutions for any of the use and occupancy classifications listed in this section, Appendix A has taken the occupancy classifications from the IBC and provided detail related to assumptions. The following sections provided some additional insight on those assumptions for the various classifications. Additional characteristics can be used for individual situations if the information is available and supportable.

A103.1.1 Assembly

These assumptions reflect nominal characteristics of persons using a public or commercial assembly space (all assembly areas outside of a one- or two-family residential occupancy) and provide the basis for such estimations as time to recognize an alarm, time to begin to exit, and time to find the way to a place of safety. This reflects the expectation that spaces where large populations are gathered will be afforded a high level of protection to avoid catastrophic losses (i.e., a large loss of life in a single space is generally perceived as being worse than the loss of one or two lives in multiple, smaller events).

A103.1.2 Business

These assumptions reflect nominal characteristics of persons using a business occupancy and provide the basis for such estimations as time to recognize an alarm, time to begin to exit and time to find the way to a place of safety. Generally, people using business spaces have limited responsibility for their own safety and are relying on the owners, managers and insurers of the space to provide an adequate level of safety. There is an expectation that large populations of the public will not be exposed to a high level of risk and that the building owners, managers and insurers retain significant responsibility for the safety of employees and visitors, but that the building or structure is not considered critical in emergency situations.

A103.1.3 Educational

This assumption reflects the fact that the people using educational spaces have limited responsibility for their own safety and are relying on the owners, managers, employees and insurers of the space to provide an adequate level of safety. There is an expectation that spaces wherein large populations of children and young adults are gathered will be afforded a high level of protection to avoid catastrophic losses (i.e., a large loss of life in a single space is generally perceived as being worse than the loss of one or two lives in multiple, smaller events) and that the building or structure may serve a necessary purpose in the event of an emergency (i.e., shelter). The tolerance for any loss in such occupancies is very low because these buildings house children and young adults, and schools play a vital role in a community.

A103.1.4 Factory—industrial

The assumptions for factory occupancies reflect the fact that the people in factory spaces have significant understanding of the risks associated with the occupancy and have significant responsibility for their own safety.

A103.1.5 Hazardous

These assumptions reflect the fact that the people in hazardous spaces have significant understanding of the risks associated with the occupancy. Although the users of the building or structure may voluntarily accept the associated risks, they have little control over the hazards and have only moderate responsibility for their own safety; thus, they rely on built-in protection to minimize the risks to the extent practicable. This reflects the perception that if the structure is classified as hazardous, the level of protection must be high.

A103.1.6 Institutional

These assumptions reflect nominal characteristics of persons in an institutional occupancy and provide the basis for such estimations as time to recognize an alarm, time to begin to exit, and time to find the way to a place of safety. These assumptions reflect the fact that occupants of institutional spaces have limited to no responsibility for their own safety and are relying heavily on the staff, owners, managers and insurers of the space to provide an adequate level of safety. In part, this reflects the expectation that spaces where large populations of confined or nonambulatory persons are gathered will be afforded a high level of protection to avoid catastrophic losses. In addition, the building or structure may play an important role in the event of an emergency. Furthermore, there is a large vulnerable population that is dependent on others. The various classifications of Group I Occupancies have a spectrum of vulnerabilities. In some cases, the occupants are physically able to react to an emergency, but their movement is restricted. In other Group I Occupancies such as hospitals, the occupants are generally free to come and go but are physically unable to undertake egress on their own. Institutional occupancies often depend on defend-in-place strategies versus building evacuation.

A103.1.7 Mercantile

This assumption reflects the fact that the people in mercantile spaces have limited responsibility for their own safety and are relying on the employees, owners, managers and insurers of the space to provide an adequate level of safety.

A103.1.8 Residential

Residential occupancies in general are buildings where occupants will be sleeping. There are different variations of types but the fact that such occupancies are in part used for sleeping makes the occupants more vulnerable from a life safety perspective.

A103.1.8.1 R-1, Transient

A103.1.8.1.1 R-1.1, Hotel/motel

These assumptions reflect nominal characteristics of persons in hotel/motel occupancies, wherein some employees are expected to be awake through the night. These assumptions provide the basis for such estimations as time to recognize an alarm, time to begin to exit and time to find the way to a place of safety. These assumptions reflect the fact that occupants of such residential spaces have limited to no responsibility for their own safety and are relying heavily on the employees, owners, managers and insurers of the space to provide an adequate level of safety. This reflects the expectation that spaces where large populations are gathered, some of whom may be disabled or impaired, will be afforded a high level of protection to avoid catastrophic losses (i.e., a large loss of life in a single space is generally perceived as being worse than the loss of one or two lives in multiple, smaller events).

A103.1.8.1.2 R-1.2, Boarding houses

These assumptions reflect nominal characteristics of persons in boarding house occupancies, wherein employees may be sleeping. These assumptions provide the basis for such estimations as time to recognize an alarm, time to begin to exit and time to find the way to a place of safety. These assumptions reflect the fact that occupants of such residential spaces have limited to no responsibility for their own safety and are relying in part on the employees, owners, managers and insurers of the space to provide an adequate level of safety. This reflects the expectation that spaces where large populations, some of whom may be disabled or impaired, are gathered will be afforded a high level of protection to avoid catastrophic losses (i.e., a large loss of life in a single space is generally perceived as being worse than the loss of one or two lives in multiple, smaller events).

A103.1.8.2 R-2, Multitenant residential

These assumptions reflect nominal characteristics of persons in an apartment-type occupancy and provide the basis for such estimations as time to recognize an alarm, time to begin to exit and time to find the way to a place of safety. These assumptions reflect the fact that the people in apartment spaces have significant responsibility for their own safety. This reflects the situation in which the owner maintains some level of responsibility in that the tenants cannot control risks in the building outside of their own living units. Where large populations are gathered and where occupants, some of whom may be disabled or impaired, are in greater density, there is an expectation that such spaces will be afforded a high level of protection to avoid catastrophic losses (i.e., a large loss of life in a single space is generally perceived as being worse than the loss of one or two lives in multiple, smaller events).

A103.1.8.3 R-3, One- and two-family residential

These assumptions reflect the fact that the people in one- and two-family residential spaces have significant responsibility for their own safety and the expectation that a person's home is his or her castle (i.e., more responsibility on the homeowner than on the local government to protect the occupants).

A103.1.8.4 R-4, Residential care

These assumptions reflect nominal characteristics of persons in residential care occupancies, wherein staff may be assumed to be asleep and provide the basis for such estimations as time to recognize an alarm, time to begin to exit and time to find the way to a place of safety. These assumptions reflect the fact that occupants of such residential spaces have limited to no responsibility for their own safety and are relying heavily on the staff, owners, managers and insurers of the space to provide an adequate level of safety. Where large populations are gathered and where occupants, some of whom may be disabled or impaired, are in greater density, there is an expectation that such spaces will be afforded a high level of protection to avoid catastrophic losses (i.e., a large loss of life in a single space is generally perceived as being worse than the loss of one or two lives in multiple, smaller events).

A103.1.9 Special use

A103.1.9.1 SP-1, Covered mall building

These assumptions reflect nominal characteristics of persons in mercantile-type occupancies and provide the basis for such estimations as time to recognize an alarm, time to begin to exit and time to find the way to a place of safety. It is generally assumed that the people in mercantile spaces have limited responsibility for their own safety and are relying on the employees, owners, managers and insurers of the space to provide an adequate level of safety. This reflects the expectation that although spaces where large populations are gathered should be afforded a higher level of protection to avoid catastrophic losses (i.e., a large loss of life in a single space is generally perceived as being worse than the loss of one or two lives in multiple, smaller events), the users of the space will be aware, awake and readily able to respond in emergency situations.

A103.1.9.2 SP-2, High-rise building

The assumed risk levels, hazard levels and occupant characteristics shall be appropriate to the uses present within the building, and the structural, fire protection and means of egress features shall be designed to accommodate the highest risk level present in the building. As a high-rise building is simply a larger and taller building of an otherwise defined use classification, the appropriate risk levels, hazard levels and occupant characteristics for the base use classification apply. Given the increased amount of people due to the larger size, the increased complexity of egress, and the increased difficulty presented to emergency responders, a higher level of risk and safety is appropriate. In multi-use high-rise buildings, the total population, the distribution of their characteristics and the distributions of the risk should be considered in selecting a base risk level. As a starting point, the highest level of risk by use group classification should be assumed. It shall also be assumed that public expectations regarding the protection afforded those occupying, visiting or working in such a building, structure or portion of the building or structure are high. This reflects the expectation that spaces where large populations are gathered, some of whom may be disabled or impaired, will be afforded a high level of protection to avoid catastrophic losses (i.e., a large loss of life in a single space is generally perceived as being worse than the loss of one or two lives in multiple, smaller events).

A103.1.9.3 SP-3, Atrium

As an atrium will be located in a building of an otherwise defined use classification, the appropriate risk levels, hazard levels and occupant characteristics for the base use classification apply. Atriums have a special category primarily related to the concerns of the vertical movement of smoke through a building during a fire. Note that atriums are essentially a vertical shaft but are given special allowances and requirements in the IBC. It should be noted that there are other ways to address floor openings in the IBC.

A103.1.9.4 SP-4, Underground building

An underground building is simply a designation for an otherwise defined use group classification that happens to be below grade, the appropriate risk levels, hazard levels and occupant characteristics for the base use classification apply. The primary reason for the special treatment of such buildings is similar to those of a high-rise building. More specifically, evacuation of occupants becomes more complex, and it increases the difficulty of managing smoke, performing search and rescue operations, and responding to an emergency.

A103.1.9.5 SP-5, Mechanical-access open parking garage

These assumptions reflect nominal characteristics of persons in a garage occupancy and provide the basis for such estimations as time to recognize an alarm, time to begin to exit and time to find the way to a place of safety. Generally, such buildings will be occupied by employees only above the ground floor who, as noted in Appendix A, will be awake, alert and able to exit without the assistance of others.

A103.1.9.6 SP-6, Ramp-access open parking garage

These assumptions reflect nominal characteristics of persons in a garage occupancy and provide the basis for such estimations as time to recognize an alarm, time to begin to exit and time to find the way to a place of safety. Unlike mechanical-access open parking garages, the general public will likely be allowed to move throughout ramp access parking garages. This will change the level of familiarity of the occupants and possibly allow a larger number of occupants.

A103.1.9.7 SP-7, Enclosed parking garage

These assumptions reflect nominal characteristics of persons in a garage occupancy and provide the basis for such estimations as time to recognize an alarm, time to begin to exit and time to find the way to a place of safety. Enclosed parking garages are those garages that cannot meet the requirements for open parking garages. Therefore, the hazards imposed upon the occupants are somewhat different. If a fire were to occur, the smoke would not dissipate as well as it would for an open parking garage.

A103.1.9.8 SP-8, Motor vehicle service station

These assumptions reflect nominal characteristics of persons in a service station occupancy and provide the basis for such estimations as time to recognize an alarm, time to begin to exit and time to find the way to a place of safety. The current code terms these as motor vehicle fuel dispensing facilities

A103.1.9.9 SP-9, Motor vehicle repair garage

These assumptions reflect nominal characteristics of persons in a motor vehicle repair occupancy and provide the basis for such estimations as time to recognize an alarm, time to begin to exit and time to find the way to a place of safety.

A103.1.9.10 SP-10, Motion picture projection room

These assumptions reflect nominal characteristics of persons in a projection room occupancy and provide the basis for such estimations as time to recognize an alarm, time to begin to exit and time to find the way to a place of safety.

A103.1.9.11 SP-11, Stages and platforms

These assumptions reflect nominal characteristics of persons using stages and platforms and provide the basis for such estimations as time to recognize an alarm, time to begin to exit and time to find the way to a place of safety.

A103.1.9.12 SP-12, Special amusement building

These assumptions reflect nominal characteristics of persons in a special amusement occupancy and provide the basis for such estimations as time to recognize an alarm, time to begin to exit and time to find the way to a place of safety. These assumptions reflect the fact that the people in special amusement buildings have limited responsibility for their own safety and are relying on the employees, owners, managers and insurers of the space to provide an adequate level of safety. There is an expectation that spaces where large populations, some of whom may be disabled or impaired, are gathered will be afforded a high level of protection to avoid catastrophic losses (i.e., a large loss of life in a single space is generally perceived as being worse than the loss of one or two lives in multiple, smaller events).

A103.1.9.13 SP-13, Aircraft-related structure

These assumptions reflect nominal characteristics of persons in aircraft-related structures and provide the basis for such estimations as time to recognize an alarm, time to begin to exit and time to find the way to a place of safety.

A103.1.10 Storage

These assumptions reflect nominal characteristics of persons in a storage occupancy and provide the basis for such estimations as time to recognize an alarm, time to begin to exit and time to find the way to a place of safety. People using storage spaces have significant responsibility for their own safety and are not relying on the owners, managers and insurers of the space to provide an adequate level of safety.

A103.1.11 Utility and miscellaneous

These assumptions reflect nominal characteristics of persons using miscellaneous occupancies and provide the basis for such estimations as time to recognize an alarm, time to begin to exit and time to find the way to a place of safety. These assumptions reflect the fact that the people using miscellaneous spaces have significant responsibility for their own safety and are not relying on the owners, managers and insurers of the space to provide an adequate level of safety.

Acceptable methods

Given that use groups are definitions, there are no acceptable methods. However, should an additional definition be required and/or a prescriptive approach to building design and construction be desired, applicable definitions and associated requirements in Chapters 3 and 4 of the IBC are permitted to be used.

Alternative classifications for building use are permitted where the use in the opinion of the designer, code official, building developer, owner or manager is not adequately defined by this code and for which adequate justification is provided.

As the use classifications are based on primary building use, occupant characteristics, risk-to-life of the occupants and importance of the building or its contents to the local community, any information that can better support any of these areas would be appropriate.

APPENDIX

B

WORKSHEET FOR ASSIGNING SPECIFIC STRUCTURES TO PERFORMANCE GROUPS

This appendix allows adjustment of the performance group based on occupants or the unique features of a building.

The table in Appendix B can be used for two purposes. If a facility is being evaluated, and there is no clear means of determining a performance group because the facility is not similar to occupancies described in Table 303.1, Appendix B can be used as a subjective means of evaluating the associated risks. From that evaluation, a performance group can be developed.

A second use of Appendix B would be to evaluate the impact of unusual features in a building or facility for which the performance group is identified in Table 303.1. For instance, if for some reason a facility that would normally be categorized as a Performance Group II is of extremely high importance to a community and contains some unusual hazards, Appendix B might be used as a means of evaluating whether the facility should be treated as a Performance Group III or IV.

The completed table in this User's Guide shows levels of risk assigned to several typical occupancies and the resultant performance group. In each case, it is a subjective analysis, and there are items in the completed table with which the user may disagree. The intent, however, is to provide a means of channeling the thought process in determining the performance group of a building or facility by reviewing the various risk factors that may be contained within the building. As more risk factors are identified, the performance group would be increased.

This table is intended to be used as a subjective analysis rather than as a numerical analysis. The user could decide to assign numerical values in this table as well. That approach, however, may result in all risk factors being inappropriately treated equally; the subjective analysis would allow the user to weight the risk factors in the analysis. Another problem with a numerical analysis is that multiple facility types would need to be evaluated in order to establish a database of relative rankings. Otherwise, the numbers may be meaningless.

WORKSHEET FOR ASSIGNING SPECIFIC STRUCTURES TO PERFORMANCE GROUPS						
RISK FACTORS	**Small Agricultural Storage Building (U)**	**15-Unit Apartment Building (R-1)**	**Outdoor Sports Stadium (A-5)**	**Large Hospital (I-2)**	**Primary Employer Factory (F-1)**	**Blank Space for Specific Structure**
Occupant Load. Maximum number of persons permitted to be in the structure or a portion of the structure.	Small	Medium	High	High	Medium	
Duration. Maximum length of time that the structure is significantly occupied.	Small	High	Medium	High	Medium	
Sleeping. Do people normally sleep in the building?	Small	High	Small	High	Small	
Occupant Familiarity. Are occupants expected to be familiar with the building layout and means of egress?	Small	Small	Medium	High	Small	
Occupant Vulnerability. What percentage of occupants, employees or visitors is considered to comprise members of a vulnerable population.	Small	Medium	Medium	High	Small	
Dependent Relationships. Is there a significant percentage of occupants or visitors who are expected to have relationships that may delay egress from the building?	Small	High	Medium	Small	Small	
HAZARD FACTORS						
Nature of the Hazard. What is the nature of the hazard, and what are its impacts on the occupants, the structure and the contents?	Small	Medium	Medium	Medium	Medium	
Internal or External Hazard. Is the hazard likely to originate internally or externally or both?	Small	Medium	Medium	Medium	Medium	
LEVEL OF IMPORTANCE						
Population. Are large numbers of people expected to be present?	Small	Small	High	Small	Medium	
Essential Facilities. Is the structure required for emergency response or post-disaster emergency treatment, utilities, communications or housing?	Small	Small	Small	High	Small	
Damage Potential. Is there significant risk of widespread and/or long-term injuries, deaths or damage possible from the failure of the structure?	Small	Small	High	High		
Community Importance. Is the structure or its use largely responsible for economic stability or other important functions of the community?	Small	Small	Medium	High	High	
SPECIFIC ADJUSTMENTS						
Are the design performance levels adequate and appropriate for the specific structure?	Small	Small	Small	Small	High	
OVERALL RISK, HAZARD, IMPORTANCE FACTORS AND PERFORMANCE GROUP	I	II	III	IV	III[1]	

1. Table 303.1 normally classifies this occupancy as Performance Group II.

(light shading)	**SMALL RISK**
(medium shading)	**MEDIUM RISK**
(dark shading)	**HIGH RISK**

APPENDIX

C INDIVIDUALLY SUBSTANTIATED DESIGN METHOD

When the design analysis and methodology are not based on authoritative documents or design guides, this method of validation may be used in lieu of code provisions. The code requires a peer review in Section 103.3.4 for any methods falling into this category.

The questions that need to be answered to determine if a design method using an unsubstantiated document is appropriate include:

- Who developed the engineering standards, design guides and standards of practice proposed for use?
- Who evaluates them on their ability to perform their intended function?
- Who validates them?
- Who deems them to be acceptable?
- How are they included in the system such that, once accepted, they can be readily used?

The individually substantiated design method was created to provide a mechanism to promote innovation but in a structured manner. An example may be the application of a new design method that was developed through a Ph.D. thesis. This method can be used once a higher level of review, perhaps a peer review and validation, has occurred relative to the appropriateness and limits of the methods application. Once a document is considered a design guide or an authoritative document, a higher level of confidence and comfort exists relative to the use of the document for design.

The design professional may use the approach of comparing the design with a compliant prescriptive code design or with a performance-based design utilizing a design guide or authoritative document to demonstrate compliance with the performance code.

APPENDIX

D QUALIFICATION CHARACTERISTICS FOR DESIGN AND REVIEW OF PERFORMANCE-BASED DESIGNS

This appendix is provided as a resource to anyone undertaking a performance-based design or review to assess qualifications of the participants performing the task. The goal of this appendix is for the design professionals, special experts and competent reviewers to have technical qualifications in education and experience associated with performance-based design. These qualification characteristics define the level of knowledge or expertise necessary to perform the functions.

Design professionals who are in responsible charge of a professional design discipline (e.g., architectural, structural, electrical, mechanical, etc.) must have the experience and expertise in performance-based design to satisfy this appendix and Section 102.3.2 of the code. Individuals on the design team within a professional discipline may supplement the design professional or principal design professional's experience to meet the requirements of this appendix.

Qualification statements must be submitted to the code official as stated in Section 102.3.2 to demonstrate compliance with this appendix. The code official has the responsibility to acquire and assign competent reviewers of performance designs and to comply with the goal of this appendix. When the code official does not have competent reviewers on staff, consultant review services should be obtained.

Peer review

Though this appendix does not specifically address peer reviewer qualifications, the following text is provided for information purposes. It is excerpted from the peer review guidelines from Section 1617A.1.41 of the 2022 *California Building Code (Title 24, Part 2)*.

SECTION 1617A.1.41—PEER REVIEW REQUIREMENTS.
[OSHPD 1, 1R, 2, 4, & 5]

*1. **General.** Independent peer review is an objective technical review by knowledgeable reviewer(s) experienced in structural design, analysis and performance issues involved. The reviewer(s) shall examine the available information on the condition of the building, basic engineering concept employed and recommendations for action.*

*2. **Timing of Independent Review.** The independent reviewer (s) shall be selected prior to initiation of substantial portion of the design and analysis work that is to be reviewed, and review shall start as soon as practical and sufficient information defining the project is available.*

*3. **Qualifications and Terms of Employment.** The reviewer shall be independent from the design and construction team.*

3.1. The reviewer(s) shall have no other involvement in the project before, during or after the review, except in a review capacity.

3.2. The reviewer shall be selected and paid by owner and shall have technical expertise similar to the project being reviewed, as determined by enforcement agent.

3.3. The reviewer (in case of review team, the chair) shall be a California-licensed structural engineer who is familiar with technical issues and regulations governing the work to be reviewed.

3.4. The reviewer shall serve through completion of the project and shall not be terminated except for failure to perform the duties specified herein. Such termination shall be in writing with copies to enforcement agent, owner and the engineer of record. When a reviewer is terminated or resigns, a qualified replacement shall be appointed within 10 working days or a timeframe mutually agreed to by the Owner, Registered Design Professional (RDP) and the Office.

*4. **Scope of Review.** Review activities shall include, where appropriate, available construction documents, design criteria, observation of the condition of structure, all new and original inspection reports, including methods of sampling, analyses prepared by the engineer of record and consultants, and the new, retrofit or repair design. Review shall include consideration of the proposed design approach, method, materials and details.*

*5. **Reports.** The reviewer(s) shall prepare a written report to the owner and responsible enforcement agent that covers all aspect of the review performed including conclusions reached by the reviewer. Report shall be issued after the schematic phase, during design development, and at the completion of construction documents, but prior to their issuance of permit. Such report shall include, at the minimum, statement of the following:*

a. Scope of engineering design peer review with limitations defined.

b. The status of the project documents at each review stage.

c. Ability of selected materials and framing systems to meet the performance criteria with given loads and configuration.

d. Degree of structural system redundancy and the deformation compatibility among structural and nonstructural elements.

e. Basic constructability of the new, retrofit or repair system.

f. Other recommendation that will be appropriate for the specific project.

g. *Presentation of the conclusions of the reviewer identifying any areas that need further review, investigation and/or clarification.*

h. *Recommendations.*

Another resource is the SFPE *Guidelines for Peer Review in the Fire Protection Design Process* (2020). It is available to download at no cost from the SFPE (http://www.sfpe.org)

APPENDIX

E USE OF COMPUTER MODELS

This appendix gives guidance regarding qualifications and information that should be provided when undertaking computer modeling. More specifically, the appendix requests that computer program data be submitted as part of the documentation. Also, limitations and applicability of the model must be included as part of the documentation. Finally, the scenarios used to run the particular model must be justified.

This is an area along with the analysis of individual performance-based designs and methods where peer review may have a role. The model is only as good as the input selected and the specific application of the model. Very often, inappropriate and unrealistic input data is used, and models often have very specific application limitations. Therefore, simply because a model has been reviewed does not mean that enough information has been provided for the specific application to be accepted. The criteria used to approve the model must be very specific in order to provide guidance to the code official or outside reviewer as to what specifically about the model has been reviewed. These criteria will provide a clear understanding of what additional review of that specific application of the model may be necessary.

REFERENCES

The following references are either documents referenced within the User's Guide or references that may be of assistance in applying this code.

ASCE 7, Minimum Design Loads and Associated Criteria for Buildings and Other Structures (ASCE, 2022)

ASHRAE Handbook for Smoke Control Engineering (ASHRAE, ICC, NFPA, SFPE 2012)

ASME A17.1/CSA B44 Safety Code for Elevators and Escalators (ASME, 2022)

ASME A17.7/CSA B44.7 Performance-Based Safety Code for Elevators and Escalators (ASME, 2007)

ASME 90.1 Safety Standard for Belt Manlifts (ASME, 2009)

ASME B20.1 Safety Standard for Conveyors and Related Equipment (ASME, 2024)

Building Department Administration,5th Edition, (ICC, 2021)

Building Safety Enhancement Guidebook [Council for Tall Buildings and Urban Habitats (CTBUH), 2002]

California Building Code, Section 1617A.1.41, Peer Review Requirements (ICC, 2022

Code Official's Guide to Performance-Based Design Review (ICC/SFPE, 2004)

Egress Design Solutions, A Guide to Evacuation and Crowd Management Planning, Tubbs, J. and Meacham, B. (Wiley, 2007)

Emergency Evacuation: Elevator Systems Guideline, Edward Cardinale, Charles Mattes, Carl Galioto, et al. (CTBUH, Chicago 2004)

Engineering Security/Protective Design for High Risk Buildings, New York City Police Department, 2009

FEMA 273, NEHRP Guidelines for the Seismic Rehabilitation of Buildings (NEHRP, 1997)

FEMA 274, NEHRP Commentary on the Guidelines for the Seismic Rehabilitation of Buildings (NEHRP, 1997)

FEMA 302—National Earthquake Hazard Reductions Program (NEHRP) Recommended Provisions for Seismic Regulations for New Buildings and Other Structures (NEHRP, 1997)

FEMA 303—National Earthquake Hazard Reductions Program (NEHRP) Commentary on the Recommended Provisions for Seismic Regulations for New Buildings and Other Structures (NEHRP, 1997)

FEMA 450, National Earthquake Hazard Reductions Program (NEHRP) Recommended Provisions for Seismic Regulations for New Buildings and Other Structures (NEHRP, 2003)

FEMA P-58, Seismic Performance Assessment of Buildings, Methodology and Implementation (January 2013)

Fire Performance Analysis for Buildings, Fitzgerald, R. and Brian J. Meacham (Wiley, 2017)*ICC A117.1, Standard for Accessible and Usable Buildings and Facilities* (ICC, 2017)

International Fire Engineering Guidelines (Australian Building Code Board, 2005)

National Building Information Model Standard, Version 1.0—Part 1 Overview, Principles, and Methodologies (National Institute of Building Sciences, 2007)

NFPA 550, Guide to the Fire Safety Concepts Tree (NFPA, 2022)

NFPA 557 Standard for Determination of Fire Loads for Use in Structural Fire Protection Design (NFPA, 2023)

Performance-Based Building Design Concepts, A Companion Document to the International Code Council Performance Code for Buildings and Facilities, Meacham, B. (ICC, 2004)

Performance-Based Seismic Engineering Guidelines, Part I, Strength Design Adaptation (SEAOC, Draft 1, revised May 5, 1998, Sections 3.7 - 3.10)

Process Safety Management Planning, 29 CFR Part 1910 (OSHA)

Recommended Guidelines for the Practice of Structural Engineering in California, second edition, October, 1995, Chapter 4.

Recommended Lateral Force Requirements and Commentary, Structural Engineers Association of California, sixth edition, 1996, Sections 104.7 and 201.

Risk Management Planning, 40 CFR Part 68 (EPA)

SARA Title III

SFPE Engineering Guide: Evaluation of the Computer Model DETACT-QS (SFPE 2002)

SFPE Engineering Guide to Performance-Based Fire Protection (NFPA and SFPE, 2007)

SFPE Fire Risk Assessment Guide, 2nd edition (SFPE 2022)

SFPE Guide to Human Behavior in Fire (SFPE 2018)

SFPE Guidelines For Peer Review In The Fire Protection Design Process (SFPE, 2020)

SFPE Handbook of Fire Protection Engineering, 5th edition (SFPE 2016)

SFPE International Handbook of Structural Fire Engineering (SFPE 2021)

SFPE S.01 Engineering Standard on Calculating Fire Exposures to Structures (SFPE 2011)

SFPE S.02 Engineering Standard on Calculation Methods to Predict the Thermal Performance of Structures and Fire Resistive Assemblies

SFPE/ICC Engineering Guide-Fire Safety for Very Tall Buildings (2021)

Structural Engineers of California Blue Book (SEAOC, 1996)

Superfund Amendments and Reauthorization Act (SARA), Title III (EPA, 1986)

Tall Building Security, Resilience & Protective Design, (Council for Tall buildings and Urban Habitat, 2019)

Vision 2000 Report (Structural Engineers of California, 1995)

ICC
INTERNATIONAL CODE COUNCIL®

Changes to the Code Development Process

The International Code Council is revising its code development process to improve the quality of code content by fostering a more in-depth vetting of code change proposals. Changes will take effect in 2024–2026 for the development of the 2027 I-Codes and will move the development process to an integrated and continuous three-year cycle.

In the new timeline:

- **Year 1** will include two Committee Action Hearings for Group A Codes
- **Year 2** will include two Committee Action Hearings for Group B Codes
- **Year 3** will be the joint Public Comment Hearings and Online Governmental Consensus Vote for both Group A and B Codes

Learn more about the new process and download the new timeline here.

23-22529

ICC DIGITAL CODES PREMIUM

codes.iccsafe.org

Go Digital and Enhance Your Code Experience

ICC's Digital Codes is the most trusted and authentic source of model codes and standards, which conveniently provides access to the latest code text from across the United States.

Unlock Exclusive Tools and Features with Premium Complete

Search over 1,400 of the latest codes and standards from your desktop or mobile device.

Organize Notes and thoughts by adding personal notes, files or videos into relevant code sections.

Share Access by configuring your license to share access and content simultaneously with others (code titles, section links, bookmarks and notes).

Advanced Search narrows down your search results with multiple filters to identify codes sections more accurately.

Saved Search allows users to save search criteria for later use without re-entering.

Copy, Paste and Print
Copy, paste, PDF or print any code section.

Highlight and Annotate any code book content to keep you organized.

Bookmark and tag any section or subsection of interest.

Stay Connected on the Go with our New Premium Mobile App

- View content offline
- Download up to 15 Codes titles
- Utilize Premium tools and features

AVAILABLE ON App Store

AVAILABLE ON Google Play

Monthly and Yearly Premium Subscriptions are available for:
International Codes® | State Codes | Standards | Commentaries

Start your FREE 14-Day Premium trial at codes.iccsafe.org/trial
Learn how to use this powerful tool at codes.iccsafe.org